INSPIRE TO INNOVATE

This page intentionally left blank

Inspire to Innovate

Management and Innovation in Asia

Arnoud De Meyer

and

Sam Garg

INSEAD
1 Ayer Rajah Avenue
Singapore 138676

First published 2005 by
PALGRAVE MACMILLAN
Houndmills, Basingstoke, Hampshire RG21 6XS and
175 Fifth Avenue, New York, N.Y. 10010
Companies and representatives throughout the world

PALGRAVE MACMILLAN is the global academic imprint of the Palgrave Macmillan division of St. Martin's Press, LLC and of Palgrave Macmillan Ltd. Macmillan® is a registered trademark in the United States, United Kingdom and other countries. Palgrave is a registered trademark in the European Union and other countries.

ISBN-13: 978 1–4039–9684–8

This book is printed on paper suitable for recycling and made from fully managed and sustained forest sources.

A catalogue record for this book is available from the British Library.

A catalog record for this book is available from the Library of Congress.

10 9 8 7 6 5 4 3 2 1
14 13 12 11 10 09 08 07 06 05

To both of my parents who passed away recently
And who always gave me the incentive to be innovative

Arnoud De Meyer

To my family and friends
And to INSEAD, a special institution so dedicated to Asia

Sam Garg

This page intentionally left blank

Contents

List of figures, tables and boxes

Preface

Our interest in innovation is an old one. One of the authors, Arnoud De Meyer, has been studying the field of innovation and R&D management for more than 25 years. When he came to Asia in the mid-1990s, he started teaching about innovation management for Asian companies. Very often he received the comment that the insights he shared with the participants in executive programmes were interesting, but that it was not always clear how they could be applied within an Asian context. A number of case studies on local companies were prepared to make the teaching livelier and to the point. Very soon it became obvious that there was enough material in these case studies to produce a short book on how to implement what we had learned in industrialized economies about innovation management in Asia. That was the trigger to start writing this book.

In the course of this project we were helped by many people. Two research associates helped Arnoud De Meyer with the first case studies and desk research. Through their initial contributions, Frederic Haspeslagh and Sachin Bhardwaj kept the project alive until Sam Garg came on board as researcher and co-author. The support of the INSEAD Foundation and Akzo Nobel in sponsoring the research is highly appreciated and without that support the project would not have been possible. All the INSEAD case studies referred to in the text are available through the European Case Clearing House (www.ecch.cranfield.ac.uk). Some of them are available on the INSEAD website – www.knowledge.insead.edu.

Many people helped us with their ideas. It is impossible to name everybody. We would like to extend our gratitude to all of you who contributed through interviews, survey responses, critical comments on presentations or suggestions for improvement to the ideas and examples in this book.

In this book we quote many examples of companies from Asia and beyond. It should be clear that what we write about them is our interpretation and does not commit these companies or their managers whatsoever. The information we provide was up to date at the time of writing the book, but may have become out of date at the moment you read the book. Please interpret the cases as interesting examples, not as the base for investment decisions.

While we were finishing the last lines of this book, we were confronted with the deadly consequences of the Tsunami of 26 December 2004. The courageous examples of some of the victims to start immediately rebuilding the infrastructure made it clear that the future will be built on creativity and action. Hope for survival tomorrow depends on innovative action today. Perhaps this book can make a small contribution to make that innovation more effective.

This page intentionally left blank

1 Innovation in Asia

Innovation is relatively new to Asia. Effective management of innovation is scarce in Asia. Until recently, innovation in Asia was a subject that commanded almost no attention. The region was booming; there were lots of investments in facilities and infrastructure. The 'tiger economies' were considered to be the new model of growth.

Much has changed since 1997 when a major financial crisis struck many countries across Asia. Since then, investment levels have recovered only partially and to a different extent in the affected countries. In Southeast Asia as a whole, they remain significantly below the pre-1997 levels up to 2003.[1] Overall, the ASEAN[2] growth rates and ASEAN 5[3] growth rates still lingered around 4.97 per cent and 4.76 per cent respectively in 2003, compared to about 7.30 per cent in 1996.[4] South Korea's 'recovery' has been worse and quite erratic. And it is important to note that the investment-led economic growth in middle and upper income economies in East Asia is unlikely to return, despite significant improvements in investor confidence and the regulatory systems.[5] The World Bank President James D. Wolfensohn commented in 2003:[6]

> Unless East Asia moves toward a technological, rather than a factor-intensive, mode of production, it will experience slower growth in the future – and with that slower growth, diminished chances to bring millions out of poverty, to give millions a chance of a better life, a better education, a better future ... It is crucial then, that the countries of East Asia, even as they struggle with the challenges of today's immediate priorities and pressing concerns, seize the initiative and begin laying the seeds for a more innovative, competitive future.

Asian companies are rising to this challenge. Data provided by the US National Science Board in 2004 indicated that patent applications from Asia are on the rise. But, as we know only too well, patents are only part of the story and a lot more is needed to create effective innovation.

Major changes are taking place in Asia which suggest that innovation will be the major engine of future economic growth for the region. On the supply side of the market, India and China are emerging as important economic players. China is becoming a major supplier of all kinds of goods to the world at a price/quality equation that is different from anybody in Asia. India is gradually becoming a major challenger too, but is currently perhaps a more important force in services. This is threatening for firms in many countries across Asia, especially at a time when these countries are forced to open their markets as a consequence of their commitments under international trade agreements.

On the demand side, the industrial customers and end-consumers have more power than ever before and are demanding extra value. This is fuelled by overcapacity in a large number of industries such as automotives, semiconductors and beer. Given that most Asian firms have little branding power, they simply tend to engage in price wars. But, for many companies, competing on low prices is a suicidal strategy in the medium or long run.

The only effective response to these supply and demand pressures is to innovate. In the following section, we explore these supply and demand elements to establish the importance of innovation. The objective is to show why innovation is important to the survival and growth of many firms in Asia today. We go on to describe how this book can help you as a manager of an Asian firm in your efforts towards innovation. To that purpose, this chapter concludes with a short description of each chapter of this book.

RACE TO THE TOP ... OR THE BOTTOM?

Imagine you are an artist who performs some interesting high-risk stunts on streets. This is a small but good business for you, and you perform your stunts in all the important locations in your town. Wherever you go, people gather around you, watch your performance with excitement, applaud you and, at the end of the show, reward you with money. Now imagine that you have two new competitors in town who perform more or less the same stunts as you do. What's worse is that wherever you want to perform at least one of them is planning to perform in the same locality on a given day. As might be expected, spectators' attention, loyalty and money are now divided. Initially, you ignore the competitors and denounce the quality of their stunts, but the spectators do not seem to agree with you. To attract the spectators back, you start performing even riskier tricks, which are again copied by those two competitors. This makes you unhappy, and eventually frustrates you so much that you stop caring about the money as such. Your objective is shifting. What you want is a *big* crowd – bigger than that for your competitors. You perform increasingly riskier tasks, many of which continue to be copied rapidly. One day, as you try your new risky stunt, hoping that your competitors do not copy it, you fall and damage your spine ... so badly that you cannot perform stunts ever again!

Now, take a few moments to reflect on the current business situation of your *real* firm. Despite a different degree of complexity, the likelihood is, in essence, that your business situation is similar to that described above. Imitation is inevitable! Often, firms tend to defend themselves by offering more and better at ever lower prices and profits erode over time. As they notice their competitors rapidly catching up, they become obsessed with them. And, as a consequence, they slowly start paying less attention to their customers and more to their competitors. Such a situation – when a firm's management is unable to take an independent path to address market needs – is often prescient of the firm's demise. In this way, despite their original intention to be on the top, firms often unknowingly exhibit strategic behaviour that makes them frontrunners in the race to the bottom!

The Chinese play

At the risk of oversimplification, it seems that China has created an environment and incentives for a race to the bottom, and this across Asia, if not the world. After the fall of Communism in the USSR and Eastern Europe, China rapidly became an interesting place for investment. This has happened especially since the famous 1992 tour of southern China by Deng Xiaoping when he urged the speeding up of economic reforms. Within a year, the foreign direct investment (FDI) more than doubled to US$27.5 billion. Following the collapse of the US stock market in 2000, firms around the world came under greater pressure to manage their costs better. China seized the opportunity to present itself as an alternative investment opportunity. It joined the World Trade Organization (WTO) in 2001. In 2002, China received US$52.7 billion, which was 8.7 per cent of the total global FDI inflows, making it the largest recipient of FDI in the world. This was a significant jump in FDI of just over US$40 billion in 1999, which was 3.7 per cent of total global FDI inflows.[7]

One could argue that attracting investments for China was made easier by the fact that the Asian financial crisis had led to a loss of confidence in other Asian economies. Increased political uncertainties and mixed degrees of economic reforms after the crisis had especially weakened those Southeast Asian competitors who had attracted significant FDI in the 1990s. A lot of what was being manufactured in Southeast Asia could now be done at a fraction of the cost and with an ever improving efficiency in the increasingly investor-friendly environment of China. By the beginning of this decade, Southeast Asia began to feel the heat as FDI, helped by all kinds of incentives, started to go to China instead.

Thus, Southeast Asian firms have been trying to move up the value chain, but it has been difficult for them to move beyond a certain stage. They moved from simple assembly in the 1960s and 70s to process improvements and developments by the early 1990s quite rapidly. At some point in the early to mid-1990s it even seemed likely that they could progress to threaten Japan,[8] but further achievements in climbing up the value chain since then have been rather limited. Despite the heavy FDI, there have been limited knowledge spillovers from foreign multinational firms to local employees and companies in Asia. Multinational corporations from Japan and the West have made use of the various incentives but have been careful to protect their technologies. For instance, given the current low-cost manufacturing threat from China, Japanese companies were being urged by the Japanese government in 2004 to keep their technology in Japan.[9] Japanese firms themselves are more cautious as they often find that their technology is copied illegally by companies in other parts of Asia who then wage price wars.

It is also difficult for them to move up the value chain because of the prevalence of protected state-owned enterprises in Southeast Asian countries and a relatively undeveloped small and medium-sized enterprise (SME) sector, (with the exception of Tawain) which in many countries across the world is often the engine of innovation. On the other hand, over time, the Chinese firms, especially the 'national champion' state-owned enterprises, have been rapidly upgrading their quality standards and technology levels. This has caused deep concerns about the 'hollowing out' of the industrial economy even in a country like South Korea, which is technologically more sophisticated than its Southeast Asian competitors.

These difficulties are hard to counter for two reasons. First, it takes vision to know where to move up to in the value chain. Second, at a minimum, it requires investment in retraining a workforce, which takes time and money. Both of these are hard to come by when one is already under threat. Thus, it is a more natural reaction to compete on prices. Many companies from the rest of East and South Asia have chosen to play exactly that game (often by themselves operating in China) even though the chances of success along that route are low. In order to gain an additional source of revenue, many firms from Southeast Asia are now becoming original equipment manufacturers (OEM) players through partnerships with Chinese companies or wholly owned operations in China. The Chinese companies, on the other hand, are beginning to develop their own brands while keeping their cost competitiveness.

One often hears the argument that a large part of the 'high-value products' coming out of China are from wholly owned foreign enterprises (usually from the US, Western Europe or Japan) or joint ventures of Chinese companies with foreign companies. But the fact that they are not Chinese companies is of limited importance, as other Asian companies still find themselves on the losing side. In any case, extensive research by our INSEAD colleagues Ming Zeng and Peter Williamson shows that 'Chinese companies are emerging to be powerful rivals – not only within China but also throughout the global markets'.[10] They argue that China's challenge to the global market has been difficult to track because there are four distinctive types of Chinese companies simultaneously challenging the global markets. First of these are the *national champions* who use their strength as domestic leaders with government support to build global brands. Examples include Haier in home appliances, Galanz in microwave ovens, China International Marine Containers in refrigerated containers and Huawei Technologies in networking equipment. The national champions are given a high degree of management autonomy, and they often have public sector, private and even foreign shareholders who are global leaders in the respective industries. Second are the *dedicated exporters* who use their economies of scale to enter foreign markets. Third are the *competitive networks* where in each network many small but specialized firms cooperate and compete to build a given category of products such as neckties, toys or Christmas decorations. Fourth are the *technology upstarts* that are using innovations developed by the government-owned research institutes in emerging sectors.

Not all East Asian countries suffer from the same level of competition with China. There is a variance in the overlap of export profile of a given Asian country with that of China. The internationalization of production chains, where a component is imported into a country and is re-exported after some processing, makes such analysis difficult due to the classification system of products. Estimates show that countries like Indonesia, Hong Kong, Thailand, Malaysia and to some extent South Korea have more overlap with the Chinese exports than a country like Singapore.[11] It is true that the quality of a given product is not the same across the region. However, it is questionable how long the current advantages of more advanced countries in the region will last. At the very least, expectations of what China could achieve should not be underestimated, given the past experiences of the West with Japan and South Korea.

The Indian effect

What China is doing to manufacturing, India is starting to do to services and R&D. Through its large, English-speaking, young and technology trained workforce, it is able to offer high-quality, low-priced services to the world. A few years ago India was known only for its software programming. Today India has a booming industry of IT-enabled services, which includes everything from back-office accounting work to telesales, remote support, medical transcription and medical testing and analysis. A large number of companies (still mainly non-Indian) in the IT and pharmaceutical industries are already doing R&D in their own centres in India, or along with their Indian partners, and are generating a good number of patents.[12] In 2004, Intel's centre in India, which was opened in 1999, emerged as Intel's largest non-manufacturing design centre outside the US and was engaged in designing a new high-end microprocessor.[13] So, whereas Chinese manufacturing is today a challenge to East Asian companies, moving towards services and more R&D is threatened by India. Almost every country in Asia feels this pressure of being 'sandwiched' to a different degree.

Demand dynamics

All this is made more critical because Asian countries have decided to open their markets under international trade agreements. Trade barriers have been lowered and governments can favour local companies only in limited ways. Significant overcapacity in many industries has created unprecedented power for the customers. As business customers face price competition themselves, they push their suppliers across Asia to provide greater value. With so much variety in consumer products, increasingly consumers are demanding better value. Better value doesn't mean low prices, although it is one possible way. Innovation is another.

Not all macroeconomic indicators are negative for Asian companies. In fact, consumer spending in Asia is rising – consumers are saving less and spending more, encouraged by cheap credit and government policies. Household consumption expenditure almost doubled in Asia in nominal terms between 1990 and 2003, and increased from 54 per cent to more than 56 per cent of total GDP.[14] But, unfortunately, many Asian companies are still providing that 'extra value' simply by dropping prices. A lack of strong brands is pushing them in this direction. But lack of innovation may leave them with no other choice. This cannot be an acceptable situation. You need to create options, *you need to innovate.*

INNOVATION: THE WAY FORWARD

Innovation is not only necessary in Asian firms, it is a great opportunity. Indeed, Asian firms are highly capable of innovating. Some of them are already world leaders in industries such as consumer electronics. Everybody has seen the successful emergence in the early 2000s of Samsung and LG from South Korea in mobile phones. They have access to high-quality, low-cost manufacturing but could price their products at the high end of the market. Overcoming this obsession with prices is possible, as we have seen with the Japan in the past. To improve the focus on innovation, recent World Bank research recommended the following:[15]

1. Invest in education quality and quantity, as well as in R&D and related innovation policies
2. Increase the efficiency in the business services sector[16]
3. Promote information and communication technologies as a means to an end – improve productivity and as a tool for international production networks
4. Enhance regional economic integration.

These steps are good ones to make the environment more suitable for innovation. But, given Asian governments' preoccupation with an infinite number of priorities, one could say, with a high degree of certainty, that it will be a long time before all these recommendations are carried out. Given the market dynamics that we highlighted as trade becomes freer, how long can Asian companies afford to wait to improve their competitiveness through innovation? Not very long! Therefore, while Asian firms should continue to demand fast action by governments in these four areas, they must immediately start taking steps *themselves.*

The arguments we developed in the previous paragraphs may suggest that we are contrasting firms from China and India with firms from the rest of the Asian countries. Let us not be mistaken. Within China and India too, there are similar contrasts. Companies in the coastal provinces in China or those based in major centres like Bangalore and Bombay find themselves in competition with companies that have lower cost structures at cheaper locations within their own country. Even within a given region in China and India, firms often try to copy the successful products, services and business models of other firms around them and engage in price wars. The same phenomenon of the race to the bottom is also taking place within India and China.

A large number of firms in Asia will have to adopt innovation as their strategy, unless, of course, they are interested in playing 'last-man-standing' kind of games. But, at this stage, Asian companies cannot yet focus on innovation driven by R&D in science and technology. Patent applications from Asia are rising. But there is not yet a sufficient scale and critical mass in R&D to produce significant commercial benefits from that. So, Asian firms, we believe, must focus on *business innovations* that address customers' needs and for which customers are willing to pay. After all, R&D by itself is of little commercial value if the products and services do not address customer needs.[17]

A second focus for Asian companies should be to *inspire confidence in customers* about products and services, generated by *Asian* companies. There are a number of areas that Asian companies can begin with. For instance, when it comes to addressing the local needs of communities with little spending capacity, Asian companies are perhaps the best placed because not only do they have local tacit and explicit knowledge and customer sympathy, but also the cost structures that firms from technologically advanced countries are unable to match. And those segments are some of the largest in the world![18]

To summarize, we have presented three points in this section. First, the supply and demand dynamics are changing in Asia, helped by the economic evolution in China and India. This makes innovation imperative for a large number of firms across Asia, including India and China. Second, we argued that, in the short and medium term, firms in Asia will have to do more business-driven innovation rather than science and technology driven innovation. And, for that, they will have to start taking steps themselves rather than wait for the governments to change policies to make the environment more suitable for innovation. Third, we highlighted the importance of inspiring confidence about Asian innovations among customers. It is the customers who make companies relevant.

WHAT CAN THIS BOOK DO FOR YOU?

The success of Asian firms depends increasingly on the quality of *management*, not only on the *economic policies*. This means that, now more than ever before, it is the managers who are the determinants of the success of their firms in Asia. This is a book for senior managers of Asian firms who want to help their firms to *innovate*. The fact that you have picked up this book and are reading it indicates that you probably have a stake in successful innovation in Asia.

Innovation in the *Asian context* is different from innovation in the developed countries of the world. This is not to say that the lessons from the US, West European and Japanese research on innovation management do not apply in Asia. In fact, a lot of them do. However, there are many specific contextual factors that make the *application* of the lessons we know about innovation management quite different. Effective innovation in Asia will require *specific* ways of managing innovation. That is the idea behind this book.

This book is based on years of observation and close interaction with senior managers in Asia. Specifically, over a period of two years, we developed a number of hypotheses about specific challenges to innovation by conducting personal interviews with senior managers in Asia and translating some of them into case studies. This work was carried out by Arnoud De Meyer, Sam Garg, Frederic Haspeslagh and Sachin Bhardwaj. It also relies on cases prepared by Chua Chei Hwee. The hypotheses were tested through a large-scale, comprehensive survey across Asia. Based on our research, we will discuss the key challenges that senior managers like yourself face when trying to innovate in Asia. Not only will you find discussion on each of the challenges, but also some of the *actions* taken by firms across Asia to *overcome* those challenges to be innovative and successful. In order to stimulate your creativity, these Asian firms have intentionally been chosen to be diverse in their size, industry and country of origin. We strongly believe that large companies can learn from small companies and the small innovative companies, despite their lack of means, can learn from the example of large companies. While we cannot prescribe any solutions for your specific firm, we hope that the numerous and intentionally diverse examples in this book will inspire you to solve the challenges that you are likely to face on your journey to innovation. Simply copying the ideas presented here will not work. The objective of the variety of contexts is to *inspire* and *stimulate* you as a manager.

In Chapter 2, we quickly review the key lessons from all the innovation management research conducted over the past few decades. We show that although most of this research has been done in the US, Western Europe and to some degree Japan, a number of lessons do apply in Asia. We briefly discuss why such lessons tell only an incomplete story for successful innovation in the Asian context. If you are familiar with the literature on innovation management, you can skim through this chapter. In this chapter, we have deliberately used examples mainly from the experience of companies in the USA, Japan and Europe, because that is where the insights came from.

In Chapter 3, we provide a summary of some of the results of a survey carried out during the research for this book. More importantly, we present in detail the findings of our research on the challenges to innovation in Asia. Some of the results may surprise you, others may not.

As a manager of a firm, in the end it is more interesting for you to know about some solutions to the challenges to innovation in Asia. We introduce you to the five broad areas of challenge, which managers in Asia could address to make their firms more innovative. (As managers cannot directly influence government policy, this is not discussed.) Rather than prescribing a solution, we present you with examples of a number of firms of different sizes in different industries across Asia to show you how different elements of each of the five broad areas of challenge can be addressed. A large number of the findings described in the book are based on personal discussions with the senior executives of these firms. This, we hope, will leave you with a sense of inspiration.

Chapter 4 is about how Asian firms need to renew themselves. We discuss how firms can unlearn their sources of competitive advantage that served them well in the past, and how they could make a considerable mental shift to learn the new key sources of competitive advantage. We also devote part of the chapter to how creativity could be fostered in your organization.

Chapter 5 is devoted to the issues of markets and marketing. Innovation starts with markets and market needs – without customers, innovation is of no value to a commercial firm. But markets for Asian innovation suffer from distance from the sophisticated markets. What a firm could do in such circumstances is illustrated. We also suggest how to make sense of innovation despite 'small and poor' Asian markets for innovative goods and services. The general lack of responsiveness of Asian customers, the lack of input during the product/service development phase, the inadequate number of early adopters and the absence of customer feedback are significant hurdles to effective innovation in Asia, made worse by the lack of reliable market data and the heterogeneity of the Asian markets. We show how a number of companies have overcome these formidable challenges.

The issue of limited resources and how it impacts innovation in Asia is discussed in Chapter 6. We specifically discuss the issue of talent and capital, while touching briefly on the basic technical infrastructure available in Asia. A wide variety of examples illustrate how companies can make up for the contextual challenges through their own perseverance and creativity.

In Chapter 7, we focus on the simple observation that getting the returns and the rents from investment in innovation is important. Being able to gain these rewards is *not* so easy in Asia. In Asia, a firm has to face the illegal imitation of products. Ineffective legal frameworks and poor implementation of the law make legal recourse difficult. Also, business models and services are copied (legally) rapidly by other competitors in Asia. Staying ahead of these illegal and legal imitators is important to profit from innovation. We present a number of different strategies and how firms have followed them to make handsome profits from their innovation efforts.

In Chapter 8, we argue that in a large number of Asian firms there is a mentality which negatively affects their innovativeness. They suffer from low self-esteem, and poor perception by customers not only in the developed markets but also within Asia. This is reinforced by the fact that Asian firms often have negative coverage in the international media. Moreover, although standard-making is so important today for success in product development, Asian companies have little influence in most of the standard-making bodies. All these elements feed into each other, making a vicious circle. Escaping this underdog status is a must for Asian companies, and without that Asian firms will not be able to control their own future. In Chapter 8, we discuss the strategies and experiences of some companies who have been able to manage this underrated challenge facing the Asian companies.

In Chapter 9, we summarize some of the inspirational examples in an overview of the areas of attention and action points that may help you and your organization to prepare for successful innovation.

We hope this book will offer you some insights and inspiration in managing your innovation efforts better in Asia. Successful innovation will translate into a better future for your company's stakeholders, including yourself. Innovation in Asia is important today. Innovation management in Asia is also different from innovation elsewhere. As a manager, you have the power, more than ever before, to take things into your hands and make your firm innovative. To assist you in your efforts is the *raison d'être* of this book.

As you will notice, the examples in Chapters 4–9 come from a wide variety of companies. Some are big, like Samsung of South Korea or Shin Satellite from Thailand. Sometimes we have chosen very small companies. One of these is Lapid Foods, a Philippino company producing and selling a local specialty called 'chicharons'. They only had five outlets when we prepared this book but still showed some interesting details on how to manage innovation. Some are low tech like Lapid Foods, some are high tech like iRiver or Neowiz, both from South Korea. iRiver is an innovator in MP3 players, while Neowiz has become very successful in the Korean market of avatars (digital cartoon characters that one buys as one's identity in cyberspace). Some are product-driven companies like Martha Tilaar from Indonesia that produces locally inspired skin treatment products. Some are service organizations like the National Library Board in Singapore, Li & Fung or Banyan Tree Hotels and Resorts. Many offer consumer products or services, but we also chose some companies that sell industrial products. Examples of these are Patkol, a Thai producer of ice-making machines, or Aapico, a producer of jigs for the automobile industry. Our choice of examples does not mean that we are convinced that all these companies will always be successful. Many of them may grow, others may merge, reorient their business or simply disappear. But at some moment in their evolution they took some managerial action that is interesting as an example for others. It is these examples that we try to capture in this book. Following these examples literally is not a guarantee for success, but they may be a source of inspiration. To *inspire you to innovate* is what we hope to achieve. Some of the examples you may know, others come from unknown companies. We deliberately looked for less commonly known examples. That is perhaps a message we also want to inspire you with: good ideas can be found in unexpected corners of the industrial space. Perhaps this book will also help you to explore these corners.

Notes

1 ASEAN Statistical Yearbook 2003.
2 The Association of South East Asian Nations was a high-growth region in the 1990s and attracted a significant level of foreign direct investment (FDI). Members, as of August 2004, are Brunei Darussalam, Cambodia, Indonesia, Laos, Malaysia, Myanmar, Philippines, Singapore, Thailand and Vietnam.
3 ASEAN 5 is a group of high-growth countries within ASEAN, comprising Indonesia, Malaysia, Philippines, Singapore and Thailand.
4 ASEAN Finance and Macroeconomic Surveillance Unit Database.
5 For a good discussion on this, see *Can East Asia Compete?* by Shahid Yusuf and Simon J. Evenett, co-published by the World Bank and Oxford University Press, 2002.
6 World Bank press release in Tokyo, 16 January, 2003, on the unveiling of *Innovative East Asia: The Future of Growth* (co-published by the World Bank and Oxford University Press). http://web.worldbank.org/WBSITE/EXTERNAL/NEWS/0,,contentMDK:20086431~menuPK:34463~pagePK:34370~piPK:34424~theSitePK:4607,00.html
7 UNCTAD World Investment Report 2004.
8 Hobday M., 1995, *Innovation in East Asia: The Challenge to Japan*, Edward Elgar, London.
9 (Still) Made in Japan, *The Economist*, 7 April, 2004.
10 Zeng M. and Williamson P.W., 2003, The Hidden Dragons, *Harvard Business Review*, **81**(10): 92–9.
11 *China's Industrial Rise*, Australian Government, Department of Foreign Affairs and Trade, Economic Analytical Unit, 2003.
12 Innovative India, *The Economist* 1 April, 2004.
13 From Intel Inside to India Inside, *Economic Times* 30 July, 2004.
14 Consumers lift Asia's growth prospects, Francesco Guerrera, FT.com site; 7 October, 2003.
15 Yusuf S. and Evenett S.J., 2002, op. cit.
16 In its research, McKinsey and Co. has often highlighted that service sectors across Asia have poor efficiency.
17 The reader may be interested in the concept of 'value innovation' as articulated by Kim W.C. and Mauborgne R., 2005, *Blue Ocean Strategy*, Harvard Business School Press, Boston.
18 For detailed discussion on this, please refer to Prahalad C.K., 2005 *The Fortune at the Bottom of the Pyramid*, Wharton School Publishing, Philadelphia. One could think about bottom of the pyramid and its needs in terms of 'disruptive technologies' as articulated by Christensen C.M., 1997, *The Innovator's Dilemma: When New Technologies Cause Great Firms to Fail*, Harvard Business School Press, Boston, and others. One could think about this generally in terms of 'value innovation' as articulated by Kim and Mauborgne.

2 What do we *already* know about innovation management?

This book is about how to innovate in Asia. We are not arguing that the fundamental concepts of innovation management are very different in Asia from those in Europe, Japan or North America. We learned through our interviews that the implementation of these concepts is influenced by the specific characteristics in the environment. This may sound obvious: implementation is always contingent on the environment. Our contribution is to go beyond a general statement and offer a more detailed analysis of these characteristics and their specific impact. But before we attempt to understand how these characteristics impact companies in Asia, we need to know a bit more about some of these innovation management concepts.

The academic literature and popular business press are full of articles on innovation management.[1] It would be a waste to spend too much time summarizing it. But we thought it would be worthwhile providing you with some ideas of what we consider essential when you want to be successful with innovation. We start with what we think is a helpful definition of innovation. And then summarize some of the literature in eight points, never to be forgotten when one innovates.

WHAT IS INNOVATION?

We define innovation as follows:

> Innovation is the *economically successful* introduction of a *new technology or a new combination of existing technologies* in order to create a *drastic change* in the *value/price relationship* offered to the *customer and/or user*.

That is a mouthful. And it sounds academic. It probably needs a bit more explanation. We italicized five elements in this definition and we now explain each of these, starting with the last one.

It is essential that innovation starts with a *customer and/or user*. From our perspective, innovation exists only if a customer or user is convinced that there has been a drastic change in what he or she perceives as value for price. We mention always both customer and/or user, because they may not be the same. In the case of an industrial product, for example, the customer is a company, but the user is the employee who does not often have any power over the purchasing decision. This also applies to consumer products. When a family buys a car, the decision about what kind of car will be selected is often made by the family head. But the users are the partner and the children, that is, the whole family. When you evaluate how the potential innovation makes a difference in the value/price relationship, you need to take into account the perception of the buyer-customer and/or user.

Innovations affect the *value/price relationship*. Remember that innovation can affect both sides of that relationship. This suggests that we will consider all forms of innovations: products, services, systems and processes. Products and services will probably have more impact on the perceived value. A new MP3 player with added benefits, larger memory or higher compatibility with other electronic products may create a new value for the user. Systems and processes will have an impact on the production cost of a product or service and this may be translated into a lower price for the same service or device.

Take the case of Dell computers. Through its process innovation in the supply chain it has been able to bring a PC, similar to those of its competitors, to the customer faster, while preserving a high level of flexibility. It is the cost-effective control over the supply chain that lets Dell achieve this. In other words, innovation can provide new or increased value, but can also offer the same value at a drastically lower cost.

Innovation is about *drastic changes*. When there is small adjustment in the value, we would not call that an innovation. It is perhaps a rejuvenation of the product or a product adaptation. Car manufacturers are masters in the rejuvenation of their models, and every three months camera producers probably offer a slightly adapted product with a few new features or a slightly adapted design. There is nothing wrong with such a product policy. These small changes are often useful to keep the market for these products active and squeeze the highest profits out of an innovation. But we do not define these as real innovations.

We are not limiting innovation to technological changes. Innovation and technology are not identical. Innovation can, of course, be the result of a *new technology*. Some of the innovations in the genetic engineering or telecommunication industry are indeed exploiting new technology. New Samsung 3G mobile phones exploiting the possibilities of GPRS, large flat-panel liquid crystal displays, for example as produced by AU Optronics in Taiwan, or genetically modified plants that enable a higher yield for the farmer are rooted in scientific and technological developments. But many of the examples in this book are based on a clever *redeployment or recombination of existing technologies*. There is nothing wrong with that either. Do you remember the first PCs in the mid-1970s, when they were still called 'home computers'? They were a clever recombination of existing TV monitors, keyboards, microprocessors and audiotape decks as external memory. Nothing really new from a technological point of view. But they were a creative breakthrough in what the concept of a computer was all about. We consider this type of creative redefinitions of the value proposition to be true innovations.

Finally, a creative breakthrough becomes an innovation when it has led to *economic success* for the firm. When innovating you need to make the distinction between technical, commercial and economic success. Technical success means that you are able to translate your dream or idea into a real product, service, system or process. This is the step of invention. It requires a lot of hard work and creativity, and is clearly a necessary condition for success in innovation. But many technically successful products have ended up on the waste dump.

In the early 1980s, a few companies introduced PCs with screens that did not have the familiar 12, 15 or 17 inch screen, but one that was identical to an A4 format. In principle it looked like a good idea to make a screen like a standard document page. That way, what you saw on the screen, you would get printed out. Creative, isn't it? Yet none of these products was successful. Remember that creativity and inventiveness are not the same as innovation. It is necessary to be creative but it is not sufficient. There is more needed to be successful as an innovator.

Many would think that an innovation is successful when you can launch it and it becomes a commercial success. Yes, this is a necessary condition. But what happens if your new product cannibalizes your previous product and does so at a lower margin? Are you better off? Obviously not. You may not have had any choice because your competitors forced you to launch the new product. But we cannot call such a commercial success, whereby you are financially worse off, a successful innovation. Many self-service food retailers who switched early on to internet-based ordering and home delivery learned this the hard way. They discovered that the consumer was not prepared to pay more for internet-based distribution. Yet the retailers had to do many more tasks themselves, for example the picking and packaging of the food and the delivery at home. Prices remained at best stable and costs went up. It was an inventive approach to distribution; for some it became a commercial success, but the margins went down. Economic success was elusive.

Therefore we will take as a condition for success that you must make more money with the innovation than with alternative investment for the resources you used to create the innovation. There is a simple way of measuring that. Many of you will have a good idea of your overall cost of capital. We argue that an innovation needs to have an internal rate of return that is significantly higher than your cost of capital in order to reward you for the risk you take with your innovation. How much higher can be debated, but we think it should be double digit percentage points. If you can realize that supplementary margin and thus have an economic success, we will call your effort an 'innovation'.

HOW DO YOU MANAGE INNOVATION?

The trouble with innovation is that, while it is ever more important for the success of your firm, it is a risky enterprise. The traditional literature exploring the subject draws discouraging conclusions: the prospects of the innovator 'making it' are slim. Overall it appears that the probability of economic success for an innovation is 20–30 per cent of all the projects you start in your organization. The implication is simple: we need to have the guts to live with risk if we want to innovate for our health.

But just living with risk is not satisfying. We also must learn how to improve our chances of success. The good news is that three decades of management studies have produced a lot of insights on how to do this. None of them guarantees success but their appropriate application may give you an edge over your competitor.

As we said in the beginning, there are hundreds of good ideas based on excellent research and the experience of many high-performing managers. We summarize these principles in eight broad categories:

1. There is no innovation without leadership
2. Innovation requires calculated risk management
3. Innovation is triggered by creativity
4. Innovation requires organizational integration
5. Success in innovation requires excellence in project management
6. Information is the crucial resource for effective innovation
7. The results of creative efforts need to be protected
8. Successful innovation is rooted in a good understanding of the market.

In the following sections we explain each of these in more detail.

There is no innovation without leadership

The first category, appropriately, is *leadership*. Successful innovation requires a clear vision defined by the leadership of the organization as well as the creation of an environment where this vision can be shared by colleagues and collaborators. Both are important. The success of Microsoft and Oracle is to a large extent influenced by the vision that Bill Gates and Larry Ellison respectively have for their organizations. Also the success of Linux is driven by the vision of those who started it, and their ability to make others enthusiastic about it. Interestingly enough, these three visions are the expression of partially divergent views on where the computer industry will go. And yet in each of the cases they rally organizations or networks and make them successful in innovation.

The leadership needs to set the goals: what kind of business the firm wants to be in and how you want to position the firm vis-à-vis the competition. By doing so, you also define what should and should not be pursued as innovation projects. A clear vision is the best way to help to define the portfolio of projects and the criteria that you need to use to evaluate new opportunities. It also helps when you need to say no: some of the most difficult decisions in innovation are precisely to say *no* to a project or stop a project that is not delivering the expected results. Ask yourself about your own organization: how many failing projects dragged on for too long, because you could not stop your enthusiastic employees working on them? Or worse, nobody dared to tell senior management that their pet project was going nowhere. Having a clear vision translated in a portfolio of projects and checking regularly whether that portfolio still makes sense helps you to weed out the bad projects.

But having a clear vision is not sufficient. Real leadership is also about ensuring that the rest of the organization has taken ownership of the goals, understands them and acts according to them. Innovative leadership requires a lot of communication, convincing and cajoling until the vision has been absorbed throughout the organization.

A good vision that can stimulate innovation should live up to two conditions: it has to combine a long-term view with concrete short-term goals and it should not be too constraining. The organization should not feel too comfortable because the challenges are defined to be far in the future. Innovation will often stretch your organization beyond its comfort zone. Stretching the organization starts today and should not be postponed into the future. It is too easy to argue that the organization needs to innovate and yet postpone that challenge 'until we have the time and the resources'. But a vision that is too constraining and too focused is not helpful either. A vision which constrains innovation to a narrow path will kill creativity and create a false sense of security because the organization knows only too well what it needs to do. Finding the right balance between these somewhat contradictory requirements about a vision for innovation, and being able to communicate it convincingly, is what is needed by the leadership for innovation.

This combination of vision and environment is what we call the *strategic context*. Organizations rely on it to harness their creativity. Without a clear strategic context, creativity may blossom, but it will be disjointed. Strategic context gives purpose and direction, benchmarks and role models. It measures progress and shows the way ahead. Take the example of L'Oreal, the successful French cosmetics company. The chairman's vision of being the world leader in cosmetics is translated into concrete goals for branding and research. In research, for example, it is translated into fairly broad and yet ambitious goals such as gaining an in-depth understanding of healthy skin and hair at a cellular level; pinpointing biological processes behind the skin ageing process, sun damage, pigmentation as well as natural hair colour, greying and loss; to synthesizing active molecules which protect, repair and colour and translating this into new products in all cosmetic fields.[2]

Calculated risk management

The second innovation challenge is to learn how to *manage risk*. All innovations will entail risk and you need to develop a system in your organization to handle that risk. This includes ensuring that the key employees of an organization act as innovators and entrepreneurs, and that they can inspire the same spirit in their collaborators. This is often described as a form of corporate entrepreneurship. All of us have some entrepreneurial talent, some of us more than others. Those who have a lot of it probably make the leap and start a venture (whether in the profit or non-profit domain). This individual entrepreneurship is not the focus of this book. Most of us need the security of a larger organization combined with stimulus to bring out the best in us. An organization is often full of people who are willing to take some risk. They just need the right context and stimuli to do so. To be a successful innovator a company will need to convince its employees to take some calculated risks.

How do you provide that stimulus? While this is dependent on the culture of the organization, it is our experience that three simple ideas can help:

1. Offer *role models* of corporate entrepreneurs who have succeeded in taking risks and are well rewarded by the organization. The role model of Steve Jobs in Apple is an extreme example but it does show how a leader, who is prepared to take personal risks, can stimulate a whole company. Don't hesitate to put forward also those corporate entrepreneurs who may have had difficulties in the beginning or may have failed with a first attempt to innovate, but who were given the benefit of the doubt by the organization. Celebrate their success.
2. Offer a *fall-back position* to potential corporate entrepreneurs: if you can reduce somewhat the personal career risk of pursuing a potential innovation, you will find more candidates who are prepared to take the plunge. If the innovation fails for a 'good' reason, that is, due to forces that go beyond those your managers or your organization can manage, you should offer the corporate entrepreneurs the opportunity to go back to their old job.
3. Put potential and current corporate entrepreneurs in *networks* with each other. Taking risks is often a lonely exercise, particularly when things don't go well. Innovators and corporate entrepreneurs need to have a sounding board to try out ideas, without the risk of losing face. They need to hear that others have difficulties too. They need to share in each other's success. All this becomes easier when you can link the entrepreneurs and create entrepreneurial networks in the organization.

Taking risks is not just about 'jumping in foolishly' though. It has to go together with assessing in a cold, calculated way what the risks are, and preparing contingencies to cope with all kinds of uncertainty. That is the second element of risk management.[3] Many organizations have developed tools or questionnaires in order to ensure that risks are known and well evaluated. And they build contingency plans to cope with the alternative outcomes of risky adventures. But any innovative project will also be confronted with unpredicted and unpredictable events. Organizations need to have systems to cope with these: flexible responses and learning from these unpredicted events are the key elements of success. Sometimes one needs to launch some experiments in parallel to try out which approach would work best. In 1991, Sun originally developed the computer language Java for graphics design in interactive television applications. This would help it to develop an intelligent universal remote control. This attempt to innovate was unsuccessful. But in a pragmatic and opportunistic move, in 1994 the Sun management saw great potential in the use of the Java language for web applications. Sun learned from its initial failure. And through its flexible response to market signals, it was able to turn a potential loser product into a big winner.

Innovation is triggered by creativity

Innovation inevitably starts with *creativity*. Many of us have some creativity and it is the organization's role to provide an environment where we *dare* to be creative. There are tools that can help us to achieve this and break out of set patterns. Many techniques have been developed by consultants and some of these do work well. In our experience and based on research, we emphasize the following five actionable ideas that can help employees to become more creative:

1. *Groups tend to be more creative than the sum of the individuals.* This is not a law of nature. There will always be a few individuals who feel constrained by groups and who are indeed so creative on their own that they are pulled down by a group. But they are really the exception. A group will only be creative if there are no political games played within the group. Groups of employees low down the organization hierarchy, which do not have huge political stakes in the success or failure of a project, will probably be more creative, while top-level groups, where the political stakes are higher (for example a board of directors), are often less creative as a whole than the sum of the individuals sitting on that board can be.
2. We also know that *exposure to information* can stimulate creativity, although information overload can stifle it. Anybody who wrestles with a problem may have lived those moments, where surfing on the net or rather purposeless browsing of literature in a library (for those among us

who still visit libraries) often leads to unexpected insights. That exposure to information helps us to become creative. But being forced to read hundreds of pages of documents in a very short time, or surfing on the net and finding thousands of relevant URLs is not stimulating creativity. Information overload may be a great tool to stop your people from being creative.

3. *Rewards and recognition* will also help, although often it seems that the recognition for the creative contributions is the more important factor of the two to stimulate your employees to come up with new insights.

4. *Pressure and deadlines* also render us more creative. Again, there is an important caveat: pressure without a clue about how to reach the goal, without a way out of a difficult situation will actually stifle creativity. Pressure will make us creative as long as escape vents are provided along the way.

5. Many consultants offer *tools and techniques* to make you more creative. Many of these tools do indeed help. Most of them are based on the observation that we are victims of our education and past experience: we see a problem and we immediately search our database of experience and lessons we learned to see how to solve the problem. In many cases it is OK to fall back on these standardized patterns: we don't want to have to be creative to tackle everyday small problems. But these patterns also hinder us when we need to be creative to solve a new problem or come up with an alternative solution. Many of the tools and techniques do help us to break down the patterns. But once we have that insight, it is often easy to develop one's own system to tear down the walls created by our experience and education. An interesting example is car development at Volvo. Recently this company asked a team of women engineers to develop a car for women. It appears that most cars are developed by men with men as customers in mind. Simply breaking that pattern led to creative insights into how to develop a different car. For example, why does everybody need a car bonnet that opens? Probably because many macho men still think they need to be able to repair the engine. But if we set aside that argument (especially now there is so much electronics under the bonnet), one comes to the conclusion that we may want to have car bonnets that only open with specialized keys used by the maintenance people. The women designing this car also paid more attention than their male colleagues to comfort and safety and in particular to getting in and out of the car: it appears that women use cars for shorter trips and do get in and out of the car more often. Also, in many cultures, women's dress code makes it more complicated to get into a car. The car has head restraints with room for ponytails and plenty of space to put handbags. Just taking a different perspective can lead to creative proposals.

Innovation requires organizational integration

Innovation is essentially an 'enterprise of the enterprise': it is a risky effort for which the whole organization should assume accountability. This *organizational integration* is a key concept for innovation. A lot has been written on the different ways of organizing: one can be functional, project oriented, networked, and so on. But the most important insight for us is that to implement innovation, the whole organization has to take ownership of it. Innovation cannot be delegated to a development group, a new business task force or a marketing department. From the first spark of inspiration to the final product, service or system, top management and operational levels have to be mobilized together if the innovation is to work.

Figure 2.1 shows the innovation process in an organization. The innovation process is split into steps or phases – idea generation, concept development, product definition, product and process development and product launch preparation. These are obviously simplified and only given as an example. The point is that any innovation process needs to be organized into such stages, with intermediate evaluation steps. These moments where one needs to take a 'go/no go' decision are known as 'gates'. This model is often described as the stage/gate model.[4]

Beneath the phases, there are four boxes representing a group in the organization: the top of the organization, the front office (marketing and aftersales service), the idea factory (research and development, design) and the back office or operational teams. Having the boxes stretching the whole

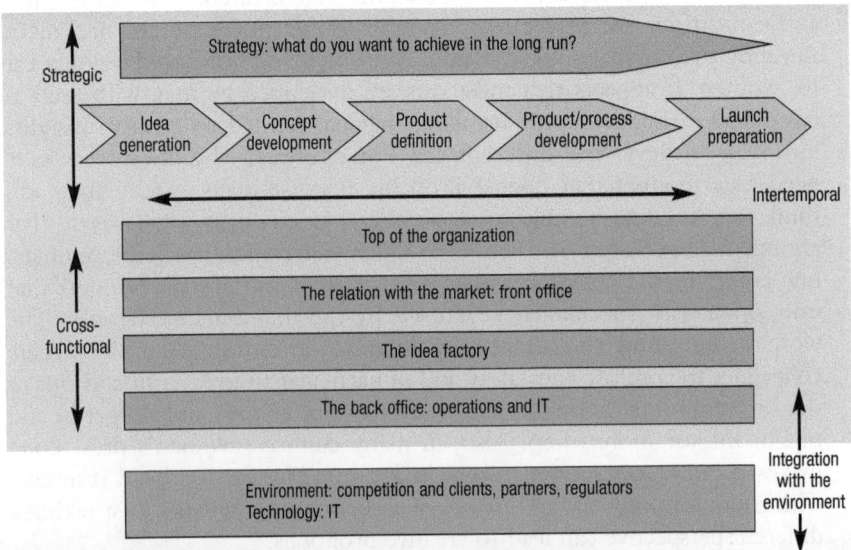

Figure 2.1 Simplified model of the innovation process
Source: Based on Angelmar[5]

spectrum from idea generation to launch indicates that there needs to be strong coordination between these different groups in the organization, from the beginning of the project till the end of the innovation effort. This does not mean that one needs project teams with a bureaucratic representation of these four groups. In fact what we need is a flexible and organic evolution of the team, always keeping in mind that all voices need to be heard throughout the innovation project.

There are two other horizontal boxes: the strategic bar representing the long-term vision of the organization and the other representing the environment, for example competition and clients, regulators and other partners such as suppliers and distributors or vendors, and so on.

Four double-sided arrows indicate the four types of integration that you will need to implement successfully when you want to innovate:

1. Integration of the project with the strategic vision of the organization: one should constantly verify that the project is aligned with and contributes to what the organization is trying to achieve in the long run.
2. Intertemporal integration between the different phases in the project: what happens at the start of a project should not hamper the activities at the end of a project. Often this has been translated into statements such as 'design for manufacturability' or 'design for serviceability'. But it boils down to the simple insight that design decisions made in the early stages of the innovation project should take into account the effect that these decisions may have on later stages in the project. Think of the cable you use for your PC. If that cable is designed such that it can easily be disconnected (as is the case for almost all of them today), you can replace it with cables that are compatible with different types of electrical plugs. On the other hand, if it goes into the black box, you would need to juggle with all kinds of adaptors in order to use your computer in different countries. It also makes manufacturing easier when disconnected cables are used: computers for different countries can be standardized and manufacturing achieves greater economies of scale. With a fixed cable, the production would have to be adjusted for each country.
3. Interfunctional integration along the project: throughout the project the different departments, functions or competence centres need to work in an integrated way.
4. Integration with the environment: the innovation needs to be compatible with the context in which it will be used. For instance, does it match the capabilities of your suppliers, can it be shipped by your distributors, are the retailers able to repair the product, is it environmentally friendly?

Organizing innovation is essentially the continuous integration of these many different actors and influences. The degree to which some actors influence innovation may differ along the way, but should never completely disappear. For example, the impact of the top of the organization can probably be more effective at the start of a project, but should never disappear. And CEOs can also play a useful role in the difficult moments during the project, or at the launch stage.

Excellence in project management

The next challenge is *implementation*. Echoing Edison and other great inventors/entrepreneurs, success in innovation requires 2 per cent inspiration and 98 per cent perspiration. Excellence in project management is the way to cope with this 98 per cent perspiration. Good project management goes beyond its application in innovation management. From that broader literature, we know that managing a project is about four major challenges:

1. Doing an excellent *planning exercise,* and putting in place the appropriate systems to *monitor deviations* from the plan and *control* for these deviations
2. Ensuring that you manage well all the *stakeholders* in the project (both those who oppose you or those who may join you in order to benefit from the innovation)
3. Organizing the *information flows:* innovation projects need information from the environment and you need to ensure that such information reaches the team; and innovation projects also need to document what they are doing, so that they can retrace their path and provide information to parties that join the project in a later stage
4. Having a *great project manager* who is sufficiently politically astute to get things done in the organization and highly enthusiastic about the end result – in the innovation literature, he/she is often called a 'product champion' and in the product development literature one talks about 'heavyweight project managers' – he or she will also have to create a motivated and capable team to execute the project.

But there is more beyond these traditional insights on the management of projects. There are two major ideas specific to innovation projects. Any recent textbook on innovation projects will tell you that there are three goals to be achieved when you develop a new product: you need to meet the design criteria and reach a high level of design quality; you need to deploy your resources in the most efficient way; and you need to deliver on time.

One major insight in the literature on time-based competition is that in innovation, time is an ever scarcer resource. To be innovative, speed in implementation must be central, although without compromising design quality and cost efficiency. Good time management is essential for innovative success. There is no point in having a great product or service that is developed within budget if you miss the optimal window for the launch of the project.

Take the example of laptops. At the end of the 1990s, laptops had a typical product life cycle in the market of six months. And every day during the life cycle the price paid for a laptop went down about one US dollar. Missing the right window to launch a product by a few weeks would drastically cut into the potential for sales and the margins one could realize. With a delay of three months, a successful project could easily be transformed into a big loser.[6] Speed of development is essential and time is often the scarcest resource in innovation. To achieve this objective of optimizing the speed at which an innovation project can be executed, concepts like simultaneous engineering, involvement of suppliers in the design of your products,[7] optimal strategies for simulation and experimentation[8] and the concept of the heavyweight project manager have been developed.

The second idea – more specific to innovation projects – is that their successful execution is heavily influenced by the uncertainty in the environment. Many good ideas have been derailed by unexpected events. Good project management in innovation needs to develop systems to cope with that uncertainty. One way to cover that is to launch projects in parallel: Honda used to start competing projects in order to get the best car. Competition between projects will stimulate creativity. But it is potentially very costly and such a strategy may only work if you can easily pull the plug on some projects before they consume too many resources. The other downside is that the teams that have to stop their projects may be disappointed. And some of your best people may leave the organization after such a sad experience. The alternative is to be flexible, to learn from the difficulties you encounter and quickly consolidate the learning.[9] Amazon seems to have been able to do this. In its still quite short existence, it has carried out lots of experiments, has been able to cut these experiments when they did not work and learn from them to prepare new offerings to its customers. Amazon today is a far cry from how it began. But its evolution has always been incremental and the company has been able to learn fast and consolidate its learning in new service offerings.

Information plays a crucial role

We have already mentioned the importance of information in project management. But the management of the information flow in innovation projects really deserves to be emphasized as a separate category.

If innovation is pictured as a production process, then information and ideas are the raw materials to be transformed into better goods or services. Access to good quality raw material is paramount in any production process. If we use that image of production to describe the innovation process, we need to ensure that the innovation team gets good access to that information. There is a whole stream of research on this subject[10] but in our experience four points deserve special attention:

1. Whatever we do in terms of making access to academic or practical literature easier, it appears that face-to-face contact is still a better way to transfer information. This remains true even when electronic communication has become pervasive. And this is particularly true in situations of high uncertainty. A lot of the information one needs during an innovation process is tacit and fuzzy and only face-to-face contact is appropriate to transfer that type of knowledge. Body language often signals uncertainty about particular decisions. No email can capture that body language convincingly. The need to enhance face-to-face communication may mean that we need to locate people together in the same physical space. If you can do this, we strongly suggest you follow that advice. But internationalization has made such centralizing and co-location more elusive. Companies often have no choice and need to locate members of a team in different geographies. And in fact globalization also has positive effects for the innovator: global networks offer other benefits in terms of ideas and tapping into expertise.[11] In today's world, marketers in Europe will collaborate with software designers in India, production plants in China and suppliers in the US. Learning how to manage these geographically dispersed teams through a better use of IT and videoconferencing in order to keep at least an impression of face-to-face communication going is a testing challenge.

2. The architecture and interior design of the buildings in which you work influence communication. This may sound trivial but the way buildings are organized has a major impact on the communication flows. The clever organization of your coffee vending machines, the strategic location of some key team players close to the normal flow of people in a building or the rearrangement of office space according to the projects, will have a positive impact on the productivity of the innovation teams. One has to be careful with some of the research in this domain, but the distances between desks beyond which face-to-face communication on professional topics drops to almost zero seem to be rather short and perhaps less than 100 metres.

3. Organization structures influence communication patterns. It is more logical to communicate within an organizational cell than outside it. No wonder that tightly knit project teams communicate well inside, but tend to lose contact with the information outside their team when they stay together too long. Designing an appropriate project structure will have to take the risk of the 'not invented here syndrome' into account: for longer projects you may want to have a functional element influencing the organizational structure.

4. Some people have a knack of understanding which information is relevant to the project and can also translate that information into the jargon used by the project team members. Think about your own organization. There are probably a few people who, whatever question you ask, seem

to know where to find information that can help you. Or if you send some of your collaborators to a trade fair, most will come back with little new information or may say they heard the same old stories again. But a few always come back with some good ideas, whatever the quality of the trade fair or conference. These people have the ability to spot information that is relevant to your organization and can translate it into the jargon you are used to. They are called 'gatekeepers'. They make a considerable and significant contribution to the improvement of the project performance. Often you find 5–10 per cent of team members have that special talent and they have to be nurtured in order to benefit from their abilities: they need to be given the space to develop their networks and gain access to information, without rendering their function too formal. Formality kills their effectiveness.

Protecting the achievements

Once you have executed your idea and produced your products or services, you will expect to reap the rewards from your efforts. Well, it's not quite that simple. Again, you must be organized. Free markets are all very well, but intellectual capital is vulnerable. You will need to *protect your innovative output.*

The logical answer to this need for protection is a combination of patents and copyrights. But patents and copyrights may not be enough. In many cases and countries the value of these legal tools is limited and enforcing them is expensive and sometimes wastes valuable time. After all, can you really prove you have been plagiarized or that your idea was the original one? And are you prepared to put your money on the table to see a court case through to the very end? You might need up to a million US dollars to pay for your lawyers. An extreme case may illustrate the difficulties of patents, even in industries where they are generally accepted. In 1960 Kilby filed a patent application for Texas Instruments for a semiconductor. The patent application has been pending for more than 30 years. During that time many companies, including some Japanese ones, used the technology and had good results selling computer equipment throughout the world. When the patent was finally granted in 1990, Fujitsu (to take an example) refused to pay royalties. This triggered a lawsuit in Japan which has been pending for a long time. This may be the stuff that large multinationals can go for, but it is probably out of the scope of a medium-sized organization.

There are other simple tools and techniques to consider: trade secrets, brands, monopolizing some input resources, trying to create captive markets, having a higher speed of development than your competitors and so on. It is probably wise not to depend on any one of these, but a combination would offer fairly good protection. Take the example of HP printers: the company has lots of patents but to a large extent it has created a protection for its technology by simply innovating faster than most of its competitors. And Mercedes can still protect itself quite well on the basis of its brand, while it also relies on patents, trade secrets and so on.

The best for the end: a good understanding of the market

Last but not least in this list of eight categories is *the market*. This is by no means an afterthought. It is our conviction that high-class communication with the customer, the user, the citizen or anybody who influences buying decisions throughout the whole innovation process is absolutely fundamental to success in innovation. A successful innovator needs good market data and lead customers who are prepared to experiment with the innovative product or service and provide feedback. A successful innovator needs to listen to how the market responds to ideas in order to quickly adjust the product or service. Downloading ring tones is the best business activity for mobile phone commerce in France. Who would have been able to predict that teenagers would be prepared to pay up to one euro to download an original ring tone? Only a quick reaction to an emerging and unpredicted demand will help the innovator in this case.

Lose contact with the clients and your magnificent breakthrough will be doomed. In a competitive world where complete originality and genius are rare, good marketing is the innovator's most important weapon for success.

SO WHAT?

What you have read is a quick summary of a vast literature. Summarizing 30 years of research and hundreds of years of experience in a chapter like this is almost impossible. But we hope it gives you a good benchmark for what is needed to be an effective innovator and to learn from some of the examples we discuss in the following chapters.

We are also pretty sure that none of it came as a total surprise. Most of the messages we gave are quite straightforward and hopefully have a high face value for you. But we wanted to give you an insight into what already exists before we ask ourselves the question: *do these simple and sound concepts apply in Asia?*

As we discuss in the next chapter, we found that these insights were indeed used by Asian managers in Asian companies. During the course of research for this book, we interviewed innovators and studied about 30 innovative Asian companies and many of them applied these lessons. But the way they *implemented* them may have been quite different. The context created additional hurdles.

How do you protect your output when the local intellectual property rights (IPR) legislation is non-existent or barely enforced? You will need to find clever ways to protect your processes or you will need to build a strong and unassailable brand. That is exactly what we saw in the case of Banyan

Tree Hotels and Resorts, a Singapore-based chain of luxury resorts. Or you may want to rely on information asymmetries, which is what Li & Fung did early on when it built a new logistics model.

How do you design products when market information is not available or incomplete? Even then you need to be able to communicate with the market. If market data companies such as Nielsen cannot help you, there is only one way out: reach out to the customers. That is what Korea's iRiver,. a pioneer of MP3 players, did when it created special shops, called iLounges, where it could study the behaviour of its 'lead' customers. And that is what Aapico, one of Thailand's automotive suppliers, did when it built close ties with the DaimlerChrysler group, so that it could learn from its principal customer.

Our basic premise is that the way we define innovation and the eight fundamental lessons about how to manage innovation also apply in East and South Asia. But we also argue that the implementation of these lessons is strewn with hurdles which are specific to the Asian environment. The cases we developed should inspire you to find ways to overcome these hurdles. Before we do that let us first spell out in the next chapter what these hurdles are.

Notes

1 Those among you who want to read more about innovation management in general can consult Shavinina L.V., 2003, *The International Handbook on Innovation*, Elsevier Science, Oxford, or Brown S.L. and Eisenhardt K.M., 1995, Product Development: Past Research, Present Findings and Future Directions, *Academy of Management Review*, **20**: 343–78.

2 See www.loreal.com.

3 If you want to have a more detailed view on how companies can organize their innovation projects in order to cope with different forms of risk, we refer you to De Meyer A., Pich M.T. and Loch C.H., 2002, Managing Project Uncertainty, From Variation to Chaos, *Sloan Management Review*, Winter, pp. 60–7.

4 See Cooper R., 2000, *Product Leadership: Creating and Launching Superior New Products,* Perseus Books, New York.

5 This model is based on R. Angelmar in the teaching note to the case Aquagliss Ultraglide, INSEAD case study, available through the European Case Clearing House.

6 This has been clearly illustrated in Crawford R. and Loch C.H., 1999, ACER Mobile Systems Unit, INSEAD case study, European Case Clearing House.

7 Clark K.B. and Fujimoto T., 1991, *Product Development Performance*, Harvard Business Press, Boston.

8 Thomke S., 2003, *Experimentation Matters*, Harvard Business Press, Boston.

9 For more information on this, see De Meyer et al. (2002), op.cit.

10 This was initiated by Allen T.J., 1977, *Managing the Flow of Technology*, MIT Press, Cambridge but see Sosa M., Eppinger S.D., Pich M.T., McKendrick D.G. and Stout S.K., 2002, Factors that Influence Technical Communication in Distributed Product Development: An Empirical Study in the Telecommunications Industry, *IEEE Transactions on Engineering Management*, **49**(1): 45–8 for a more recent overview.

11 More can learned about this from Doz Y., Santos J. and Williamson P,. 2002, *From Global to Metanational: How Companies Win in the Knowledge Economy*, Harvard Business School Press, Boston.

3 What is different about the implementation of innovation management in Asia?

The core thesis we develop in this book can be summarized as follows: The innovation management lessons described in the previous chapter apply in Asia as well as they do in the rest of the world. But their implementation needs to take into account the specific hurdles that exist in non-Japan Asia.

How can we be so sure about this? In this chapter we explain some of our research methods and results. The results of this research will provide insights into the specific hurdles for the implementation of innovation management lessons in Asia. And in the following chapters, we offer some inspiring examples of how companies have overcome these hurdles.

How did we proceed? In the first phase of the research we collected information about innovation management in Asia through interviews with privileged observers (for example scholars, leaders of innovative firms and policy makers). We collected files on about 30 innovative firms or groups of companies.[1] Based on this series of case files we developed five categories of hurdles for innovators in Asia. In the second phase these five categories were translated in 32 statements. These statements were sent in the form of a survey questionnaire to senior managers operating in Asia to get their opinion on how they influenced, positively or negatively, the practice of innovation management in Asia. The detailed results are reported elsewhere.[2] Here we present only the most important results that may help you in analysing your organization's situation concerning innovation management.

WHAT DID WE LEARN FROM THE CASE FILES?

Before we address some of the preliminary results of the exploratory study, we need to make an important caveat. Our research did not cover breakthrough innovations, neither through the case studies nor through the desk research. This does not mean that there are no breakthroughs in Asia but only that we did not have any in our sample. In most of the cases we observed opportunistic adaptations of products or services to the local markets, or exploitation of local market advantages. For example, we studied the case of Banyan Tree Hotels and Resorts,[3] a Singapore-based chain of hotels and resorts. It would be difficult to argue that such a chain of luxurious and well-located holiday resorts is a breakthrough innovation. But this company has redefined completely what the customer will experience during a weekend in one of its resorts. This redefinition puts the emphasis on individual experience, under the aptly chosen slogan 'a sanctuary for the senses'. It takes advantage of the quite exquisite physical environment of some Asian islands, as well as the image of personalized service that is often identified with Asia. The company is offering a new value proposition in the way our colleagues Kim and Mauborgne define value innovation.[4]

As we suggested in Chapter 2, many of the lessons we learned in the past 30 years about innovation management do apply in Asia. For example, the case on MyWeb.com,[5] a small producer of electronic set-top boxes which enabled access to the internet via TV monitors, is an almost classic story of an entrepreneurial venture and the company's final demise was largely due to the same reasons that start-ups go under anywhere in the world.

But let's focus here on what the five categories of additional hurdles can be (Box 3.1.).

Box 3.1 The five categories of hurdles of implementing Western innovation management lessons in Asia

1. The resources needed for innovation (in particular, technical expertise and risk capital) are still scarce.
2. Markets that stimulate innovation are geographically and/or culturally far away and, if they exist locally, are too small.
3. Existing industrial policies are aimed at catching up with the industrialized world, rather than seeking value creation through innovation.
4. Many organizations have innovation-averse organizational cultures: as a consequence there is self-fulfilling prophecy of an underdog mentality and a hierarchical organization that hinders creativity.
5. There is a considerable lack of appreciation for intangibles in Asia.

The resources needed for innovation (in particular technical expertise and risk capital) are still scarce

Contrary to what one may imagine when faced with the absolute number of engineers and scientists in China, India or Southeast Asia, there is quite a dearth of capable technologists in many of these countries. Table 3.1 shows the number of engineers and scientists in R&D per million people based on data provided by the World Bank for 2002. It is noticeable that the number

Table 3.1 Number of engineers per million persons in selected countries

Country	Scientists and engineers in R&D per million
China	545
Japan	5095
South Korea	2318
Singapore	4140
Hong Kong	93
Thailand	74
Indonesia	na
India	157
USA	4099

Source: World Bank (2003): www.worldbank.org/data/

of engineers in R&D, while large in absolute terms, is still low in proportion to the total population in countries like China, Thailand, Indonesia and India when compared with Japan or the US. Even in South Korea, there is a significanly lower number of engineers compared to these two countries. While we do not have hard data on the qualifications of the technical staff, we got the impression through the interviews that companies in Asia will be confronted in the future with both a shortage of experienced employees and a lack of quality in training.

One may argue that each year the universities in China or India produce hundreds of thousands of new engineers and scientists. But compared to the size of the country this is probably insufficient. These waves of bright people flowing out of the universities will definitely contribute to the creation of innovations in Asia, but overall there probably will remain a scarcity of experienced technical experts.

Engineering capability is only one aspect of resources. Finance is another. Several times we were told that the Asian financial markets simply lack the sophistication and willingness to invest in innovation. Risk management is not one of the strengths of Asian financial institutions and the consequence of this is a lower propensity to invest in innovative projects. An example may illustrate this. In one of our cases we were told that as a result of the risk aversion of Asian financial institutions, the cost of capital could be about 1–1.5 per cent higher than for a similar project in the US. Asian providers of capital have less experience with risk management and charge higher for the risk they take. US-based providers of capital master the tools and techniques of risk management much better but charge a risk premium to Asian companies for the uncertainty created by the geographical distance or the perceived instability in the region. While this difference of 1–1.5 per cent may seem low, it can make a significant difference in international competition.

Beyond the human and the financial resources, we discovered two resources related to what one may call an inadequate mindset. While a mindset is not a resource in the traditional sense of the word, the right mindset is sine qua non for successful innovation and can thus be considered to be a kind of intangible resource.

First, many of the managers in Asia have a clear mindset that drives them in the direction of cost reduction, as opposed to new value creation. In particular this is true for those that have been successful in the past. They often perceive low labour cost and access to cheap natural resources as the cornerstone of their company's competitive advantage. Making the mental switch from being a receiver of technology and a producer for a principal in the industrialized world to assuming the role of a proactive innovator for local and international markets is not an obvious move for many of them.

Secondly, many Asian managers still think too often in terms of product maps as opposed to process and process capability maps. Subcontractors are used to getting blueprints of products they need to build. Their business model is based on recombining and producing existing products and components. Innovators often have to recombine their process capabilities in order to design and develop new solutions for new needs. The difference between the early chip factories in Singapore and Taiwan illustrates this. In Singapore, the chip-making companies bought turnkey factories but there was little local knowledge about the intricacies of wafer production. In Taiwan the wafer production was often developed by engineers who had spent the better part of their professional life in the USA working for US companies and had developed an intimate knowledge about the production processes. The first group thought in terms of products and off-the-shelf technology. The second group had a clear map of the process capabilities needed to have a continuous evolution of wafer production. It is the second group that has been more successful in the short and medium term in rolling out innovations.

Markets that stimulate innovation are geographically and/or culturally far away and if they exist locally, are too small

The eighth lesson of innovation management, described in Chapter 2, is that we need outstanding marketing. There is no reason why this should not be true in an Asian context. But we found that the application of good marketing concepts becomes a challenge, because of the geographical distance from sophisticated and trendsetting markets and the lack of reliable market data. More specifically we found the following:

- The average Asian company has a limited understanding of, or experience with, marketing. Many of the companies in our sample, whether the automotive producer Aapico, Shin Satellite in Thailand or Biocon in India,[6] had excellent technological capabilities but little knowledge about brand building, developing sophisticated distribution channels or even advertising.

- A substantial disadvantage for innovation by Asian companies is that they are far removed from the sophisticated consumer markets in industrialized countries. Tokyo, which is often the trendsetting market for Asia, the US and some of the leading European markets, are all either geographically or culturally very distant from Southeast Asia. Asian companies need to have an extremely long and sophisticated marketing arm to be able to tap into the intimate market knowledge that is needed to develop new products or services.

- There are sophisticated local markets in development. Singapore, Hong Kong or Taipei offer lots of opportunities for designers. 'Hand phones' (the local English for a mobile telephone) are indeed that: take a short walk through the streets of Asian capitals and you will see hundreds of young and not so young texting on their mobile phones, sometimes at amazing speeds. Manila was in 2003 by far the most sophisticated market in the world for SMS as we described in the case Pinoy2Pinoy.[7] In 2005 the number of SMS in China is expected to be 550 billion. But these markets are often still embryonic and not mature enough to be a source for products that can be rolled out worldwide.

- There is a serious lack of reliable market data in Asia. In many cases it is just not there. In other cases it may be available, but no one trusts it. The lack of accuracy of market data and the simplistic way of collecting it in China has been well documented. We found that most of the time companies had no good market data available.

- On top of this lack of market data many of the Asian markets are still quite closed, if not parochial. There are historical reasons for that. Many Asian nations have had a less than happy relationship with each other in the past or are culturally quite far apart from each other. The differences between Buddhist Thailand and Muslim Malaysia are quite substantial. And the historic differences between China and Japan still influence the buying patterns in both countries.[8] This makes it more difficult to develop Asian products for Asian-wide markets. And existing market research is not easily transferable from one country to another.

Existing industrial policies are aimed at catching up with the industrialized world, rather than seeking value creation through innovation

Virtually all Asian governments have played and still play a strong role in the development of the local economy. In the past governments have chosen sectors in which they wanted their national champions to be present. The example of Malaysia trying to carve out a position in the automobile market through the Proton venture is a typical one. But it is far from unique. More often than not the governments have an important stake in the leading companies, either through direct participation in the equity or by having a strong voice in their management. This may have been an appropriate policy for economies that needed to catch up, or for companies that were acting as a subcontractor for principals in other parts of the world. But when it comes to driving innovation this is less appropriate. Nowhere in the world do governments have a particularly good track record in choosing winners in terms of innovation.

If the innovative companies *are* independent from the government, they are confronted with additional hurdles:

- The regulatory environment conducive to the growth of innovation and entrepreneurship is often lacking: innovators need good protection of their intellectual property rights (IPR) and in particular enforcement of these rights; they also need legislation that enables the fast creation of companies, but if the innovation is not successful, equally they need legislation that allows them to stop the business in an efficient and clean way.
- In many Asian countries the public sector is still one of the most important purchasers. Local governments tend to be conservative in their procurement and do not favour local innovators over well-established international brands.
- Pro-business policies and regulations have often favoured local entrepreneurs who took advantage of information asymmetry through their good contacts with the government: they did not need to innovate, it was often enough simply to set up shop and create jobs. Favouring insiders with trade protection and insider information does not necessarily lead to innovative developments.

Many organizations have innovation-averse organizational cultures

Some aspects associated with Asia or Asian cultures impede innovation or make it more difficult to penetrate new markets with Asian innovations. First there is the structure of local companies. We do not want to make sweeping generalizations, but we have observed that many firms in Asia are either family owned or have a strong family culture. This leads to a certain level of authoritarianism, and organizations with subservient and sometimes disengaged employees. Often authority is delegated to a limited extent and the top, wanting to keep control over the firm, becomes a bottleneck in decision making and the stimulation of creativity.

Secondly, Asian goods were often perceived in industrialized countries to be cheap goods. Asian factories were often seen as suitable for low-cost but also low-value work. This perception is still prevalent and in the interviews we heard several anecdotes about how senior managers in European head-quarters would confuse the manufacturing capabilities of a Thai or Malaysian company with the naive perceptions these managers gained from holiday experiences they had enjoyed in those countries.

This perceived low quality in export markets is sometimes reinforced by the poor self-image of Asian companies. It may be a somewhat irrelevant example but in some Asian capitals the mannequins used in fashion houses have Caucasian traits (barring those countries where this is explicitly forbidden). This is perhaps a trivial example of how Asian companies still don't see their own markets as mature and leading.

It has been pointed out to us that part of the organizational culture that had a negative effect on the urge to innovate had to do with the traditional attitude of the Chinese towards innovation. It is true that deep in history the Chinese Empire contributed a lot to innovation (from book printing and irrigation techniques to explosives). But most of the time these inventions were solely for the benefit of the emperor and the courts. It took European entrepreneurship to turn book printing into an international commercial success. The question can be raised whether this earlier attitude still has an impact on the attitude of today's entrepreneurs towards innovation. We have no answer to this question but the hypothesis that there is something in the Chinese culture that limits the wish to innovate deserves more research.

Finally, there is often a misunderstanding of what entrepreneurship really is. We were often told how great and strong the entrepreneurship is in the Chinese business world. But a careful examination of many success stories about Chinese entrepreneurship reveals that in fact these are success stories about trading, exploiting information assymetry and property (land) deals. There is nothing wrong with these activities, but they are rarely about value creation through innovation.

Lack of appreciation for intangibles

In our discussions and case studies we observed a strong lack of appreciation for the intangible side of innovation: brand building is often neglected, or reduced to finding a cute name; lip service is paid to the protection of IPR, and nobody seems to care about the copying of software or other intangible content. While there are good Asian brands outside Japan, with the exception of Samsung, none have made it to the top 100 brands in the world.[9]

Only hard tangible products seem to have real value in many business-people's perception. Perhaps it has something to do with the fact that until now Asian companies had a strong negative trade balance concerning IPR. They had to pay the royalties and the licences and they did not receive a lot.

Appreciation for intangibles may rise the moment these companies can gain some benefit from it. In fact that is what you see happening today in China and Thailand. The consequence today of this past attitude is that local innovators hesitate to innovate – they do not get returns on their investment in R&D as they have limited means of protecting their innovations.

A second aspect of this lack of appreciation for intangibles is the absence of good design capabilities. Asian companies may have started to invest in development, carry out some marketing, even apply some of the more sophisticated finance techniques, but often do not have the design capabilities that help to set a product apart. Japan is currently developing as a design powerhouse in Asia and is determining the fashion and technical trends. But in the rest of Asia, be it Seoul, Singapore, Hong Kong or Shanghai, there is no critical mass in good designers, whether within the companies or on a freelance basis.

CONFIRMING THESE HYPOTHESES THROUGH A LARGE SURVEY

The five categories of observations that we formulated on the basis of our interviews led to an abundance of potential hypotheses about what really hinders innovation management in Asia. Handling a multitude of qualitative observations becomes unwieldy. It does not allow us to say which one is the most important, or whether some hurdles were more important for some groups and less for others. Therefore we decided to test some of these through a large-scale survey.

The results of the interviews were translated in 32 statements about the key success factors that would affect innovation management in a positive or negative way (see Box 3.2 for the full list). This was combined with a few questions about the company and the respondent as well as a set of questions about the changes in the competitive environment and the need to innovate. The respondents were also given an opportunity to choose the top three challenges amongst those highlighted in the 32 statements and highlight any other challenges not covered in the 32 statements. Through a questionnaire, we collected the opinion of 336 senior managers operating in Asia on the importance of each of the statements for innovation in Asia. The composition of the sample is described in Table 3.2. If you want to know more about our methods, you can find them in Box 3.3.

Box 3.2 List of statements used in the survey

1. Disengaged and subservient employees
2. Overreliance on creative improvisations
3. Insufficient project management capabilities
4. Greater emphasis on market share than on profitability
5. Quality of competitive intelligence

---

cont'd

6. Stigma associated with failure
7. Quality of:
   a. engineers
   b. designers
   c. managers
8. Inadequate risk capital
9. Few role models of successful innovative companies in the country or region
10. Quick imitation of innovative business models/products/services by competitors
11. Inapplicability of innovative management lessons from the West
12. Strong cost reduction attitude
13. Lack of pressure from financial markets
14. Unsophisticated existing customer base
15. Asian markets are either too small or too heterogeneous for profitable innovation
16. Lack of reliable marketing data
17. Inadequate access to a significant number of 'early adopter' type of customers
18. Asian customers perceive Western goods to be better than Asian ones
19. Prospects of good growth for the company even without innovation
20. Inability to recombine/reconfigure existing capabilities into new products/services
21. Western markets look down on Asian products and services
22. Lack of involvement of Asian companies in setting up global standards
23. Lack of self-confidence of Asian employees in international business
24. Geographical distance from Western markets makes it difficult to understand their needs
25. Mainstream international business media reporting in English tends to focus on negative Asian business news
26. Government intervention in business in the home country of your company
27. Strong and conservative public sector leaves few opportunities for private sector
28. Inadequate protection of intellectual property rights
29. Lack of strong brands
30. Conservative business partners
31. Lack of diversity in the workforce
32. Insufficient attention to detail

Box 3.3 How did we analyse the survey data?

The questionnaire was made available through the internet and invitations to answer the questionnaire were sent by email to about 3160 senior managers, who had been in contact with INSEAD over the past few years either through executive programmes or our research activities. We received 336 usable responses. This is a response rate of 10.5 per cent. The composition of the sample is described in Table 3.2.

We want to highlight two shortcomings of the sample. While it is reasonably diverse in terms of geography, there is perhaps an overrepresentation of respondents located in Singapore. But one should take into account that in many cases Singapore acts as a hub for the region. Many regional headquarters are located in the city state. When a researcher wants senior managers to respond, one automatically ends up with many responses from a Singapore location.

Secondly, any questionnaire suffers from self-selection by the respondents. All the respondents are of the opinion that innovation is becoming significantly more important in the competition than it used to be five years ago. We probably received proportionally more responses

cont'd

from managers interested in innovation than from those for whom innovation is still irrelevant. For our study this was an advantage rather than a disadvantage: we may assume that our respondents are more acutely aware of the hurdles and advantages for innovation in Southeast Asia.

For each of the 32 items and the questions regarding the intensity of competition and the importance of innovation, we asked the respondents to score on a scale from 1 (very negative effect on the innovation in your company) to 5 (very positive effect on the innovation in your company). We also asked the respondents to indicate their top three challenges. Tables 3.4a and 3.4b record the top and bottom items for the two analyses.

Handling the responses to a list of 32 items is a difficult exercise. In order to get a better grip on the data, we applied two straightforward multivariate techniques. We reduced the number of 32 key influencing factors to a set of 10 factors. These can be interpreted as 10 dimensions along which the senior managers who participated in the survey differed. Once we had these 10 dimensions, we applied a cluster analysis to find four clusters of respondents with common characteristics. These four clusters can be seen as four groups of managers who have a common view on the problems they face with respect to innovation.

Table 3.2 Sample composition

Descriptor	Percentage (absolute number)
Company headquarters (n = 336)	
Asian	57.4
Non Asian	42.5
Title of respondent (n = 336)	
President, CEO, managing director	35.2
General management	29.2
Functional management	11.9
No answer	23.5
Industry composition (n = 335)	
Industry	48.8
Services	51.1
Geographical composition (n = 336)	
(of the location of the respondent)	
Singapore	34.2
Mainland China	10.4
Hong Kong	8.9
Malaysia	7.1
India	6.5
South Korea	4.5
Thailand	3.0
Taiwan	2.7
Philippines	2.1
Sri Lanka	0.9
Pakistan	0.9
Others	2.7
Undetermined	16.1

RESULTS FROM THE SURVEY

Importance of innovation to the respondents

As an introduction to the questionnaire we asked a few questions about the current importance of innovation compared to the past and the changing competitive environment (Table 3.3). Interpreting the results is not difficult: the respondents argue that innovation is frequent in their industry, the intensity of the competition is significantly rising and innovation is currently significantly higher than it used to be five years ago. It was interesting to see that the three indicators are significantly correlated with each other. In other words, a higher intensity of competition leads to a stronger need to innovate and is also translated in a higher frequency of innovation. But it could also be seen that the frequency of innovation lags behind the intensity and importance of innovation, perhaps indicating the challenges of innovation.

Table 3.3 Perceived changes in the environment compared to five years ago

Rating	Intensity of the competition %	Current importance of innovation %	Rating	Frequency of innovation in the industry %
Significantly higher	73	56	Very frequent	18
Higher	22	35	Frequent	42
No difference	4	9	Regularly	34
Lower	2	1	Occasionally	6
Significantly lower	0	0	Almost never	0
Mean	1.35	1.54		2.29

Hurdles and success factors for innovation in Asia

Tables 3.4a and 3.4b show the statements that describe the most positive and most negative influences on the management of innovation in Asia. Table 3.4a shows the ranking based on the individual scores for each of the statements. Table 3.4b shows the ranking on the basis of the answer to the question: which of these factors do you see as the most challenging?

As one would expect, the two tables have some statements in common and some differences. Which factors have a negative impact on the ability to innovate? Insufficient project management capabilities, the inability to reconfigure the existing capabilities of the company into new products or services and inadequate IPR protection stand out in both types of analysis. They are complemented with the negative impact of quick imitation of innovative products by competitors, weaknesses in marketing, lack of employee engagement for innovation and insufficient risk capital.

And what is not a problem? The respondents seem to have fewer problems with the intrinsic quality of the people (and score themselves as managers as

Table 3.4a Most positive and negative factors based on the individual scores of each item (on a scale from 1 to 5)

Most positive factors (or not hindering innovation)	Most negative factors (or perceived to be a hurdle)
Quality of managers	Quick imitation of innovative products by competitors
Quality of engineers	Inadequate protection of IPR
Quality of designers	Insufficient project management capabilities
Quality of competitive intelligence	Inability to reconfigure existing capabilities into new products
Asian customers perceive Western goods to be better than Asian ones	Unsophisticated existing customer base
Strong cost reduction attitude	Lack of reliable marketing data

Table 3.4b Most and least often cited factors as top three challenges for innovation (n = 290)*

Least often cited top challenges	Most often cited top challenges
Mainstream international business media focus on negative news from Asia	Disengaged employees
	Strong cost reduction attitude
Geographical distance from Western markets	Insufficient project management capabilities
Asian consumers perceive Western goods to be better than Asian ones	Inability to reconfigure existing capabilities into new products
Lack of pressure from financial markets	Inadequate IPR protection
Western markets look down on Asian goods	Inadequate risk capital

* The number of respondents is slightly lower than for the other questions. Not all respondents filled out the question about these three challenges

a strong positive factor) and don't seem to think that some of the statements that reflect an underdog mentality (for example negative reports by the business press, negative perceptions of Asian versus Western goods or distance from the markets) have a negative impact on the innovation process.

Something that seems difficult to explain is the ambiguous impact of a strong cost reduction attitude, which does not appear to be a negative factor when one looks at the individual scores of the 32 statements, but is cited among the top three hurdles to overcome. The reason is simple: the score on this item is strongly bimodal, that is, some respondents consider it to be negligible while others see this as a very important hurdle for innovation. The same respondents who score it as an important hurdle also mention it consistently as one of the most important challenges.

We originally thought that the positive or negative impact of key factors for innovation would be perceived differently by managers depending on whether their companies have headquarters inside or outside Asia (further described as Asian and non-Asian companies). But we were wrong. There are not that many statements for which we find significant differences. We find only five items on which managers of Asian companies are significantly more positive than their counterparts of non-Asian companies:

1. The negative impact of insufficient project management capabilities
2. The degree to which Western management methods can be applied in Asia
3. The negative influence of a cost reduction attitude
4. The lack of reliable market data
5. The negative impact of inadequate IPR protection.

It does appear that the managers of non-Asian companies are a bit more negative about the management capabilities of their staff or Asian management in general. But all in all the number of significant differences is not high.

Understanding how senior managers differ in their opinion

Interpreting a list of 32 statements remains a laborious exercise and therefore we reduced them through an appropriate statistical technique to the underlying factors that describe how the respondents have a different view on the difficulties of implementing innovation in Asia. We found 10 underlying factors that explain the differences in opinion in our sample of 336 senior managers. In the order of decreasing importance, they are as follows:

1. *The absence of an environment in Asia in which it is easy to operate as an innovator.* This combines the lack of market data and trendsetting customers with the stigma associated with failure. These factors don't make it easy for some innovators to pursue their projects, while others don't seem to be bothered by it.
2. *The underdog mentality* of the Asian company: to what extent are Asian companies involved in setting international standards, do they lack confidence, do have they good market perception and do they see the international business press as constantly negative about Asia? Again some managers see this as a real hurdle, while others don't seem to be affected.
3. *The lack of some knowledge resources,* for example the lack of high-quality knowledge workers and reliable competitive intelligence.
4. *The inertia created by the forces of tradition in Asian business.*
5. *The lack of basic management models and lessons specifically applicable to innovation management in Asia.*
6. *The negative impact of government.*
7. *No perceived need to innovate.*
8. *The lack of external pressure and business rewards for innovation.*
9. *The lack of good market understanding.*
10. *The traditional cost reduction attitude.*

Once again we wanted to know whether the significant differences observed in the original 32 items between respondents from Asian and non-Asian companies were confirmed in the 10 factors. There is only one factor where there is a real difference: managers of non-Asian firms felt that the perceived need to innovate was lower than managers of Asian firms did. The important observation seems to us, however, that on the factors that explain the highest variance, there is no difference between Asian and non-Asian managers.

Classifying the sample into groups

Having 10 factors that explain the differences between the respondents, we can now explore to what extent the group of 336 respondents could be grouped into groups with common characteristics. We found four distinct groups with strong common characteristics (Table 3.5).

Managers in the first group consider the lack of appropriate management methods and the lack of a perceived need to innovate to be important hurdles. But on the other hand they think that the underdog mentality, the inertia due to the traditional Asian approach to management and the lack of external rewards for innovation do not hinder them too much. This may suggest that this is a cluster of managers who see the major challenge for innovation in Asia as the need to get incentives to innovate. They also require the development of appropriate innovation management methods. They can perhaps be seen as people who want to innovate but don't know how and seem to be still at the beginning of the innovation journey. We will call them *innovation starters.*

The respondents in the second cluster feel positive concerning the support that the economic environment can provide for innovation. They have no problems with the availability of knowledge resources or the lack of a perceived need to innovate. On the other hand, they feel somewhat negative

Table 3.5 Innovation types

Type	Have trouble with	Have less problems with
Innovation starters	Appropriate management methods for innovation Lack of a perceived need to innovate	Underdog mentality Inertia due to traditional Asian management style Lack of external rewards for innovation
Tradition fighters	Underdog mentality Inertia due to traditional Asian management style	Availability of knowledge resources Lack of a perceived need to innovate
Poor in knowledge resources	Availability of knowledge resources	Appropriate management methods for innovation
Stuck in the muck	External rewards for innovation Influence of the government Underdog mentality	Availability of knowledge resources Appropriate management methods for innovation Cost reduction attitude

about the underdog mentality and negative about the inertia created by the traditional Asian approach to management and the typical cost reduction attitude. They seem to believe one can easily innovate in Asia if one can overcome some of the traditional behavioural hurdles for innovation in Asia. They may well be those managers who are already innovating but feel they are constrained by some administrative heritage from the past, both in their firm and the environment. We will call this group the *tradition fighters*.

The third cluster scores average on most factors except two. They have no problems with the lack of appropriate management models, but they strongly feel that they do not have the right resources available in the firm. This group perceives themselves to be poor in the knowledge competencies needed for innovation. We call the respondents in this group the *poor in knowledge resources*.

The fourth cluster seems to believe that they have the resources, understand the methods and do not suffer too much from a cost reduction attitude. But they feel they need improvement on the innovative economic environment and a reduction in the underdog mentality. They believe that innovation is insufficiently rewarded both in product and financial markets. They feel also negative about the influence of the government. In summary they believe they can innovate but require the general environment to improve. On some items they are the opposite of the first group. In one sense, they are convinced they should and can innovate but blame the environment for creating hurdles. They are like a strong truck that can and wants to move but is stuck in the mud of the environment. Therefore we call this group *stuck in the muck*.

Once again, we analysed these clusters to see if they were influenced by the composition of the sample and found that whether the respondent came from a company with an Asian or non-Asian headquarters did not influence the innovation type.

WHAT DO WE DO WITH THESE RESULTS?

We would like to summarize our results of the survey research under four headings:

1. We can derive a set of factors that are specific to Asia and that influence the implementation of innovation management in Asia
2. Not all our respondents have a similar opinion and there are 10 consolidated factors on which managers in Asia differ in their opinion about innovation management
3. In the interpretation of these factors there are only minor differences between managers from Asian and non-Asian companies
4. Not all companies/respondents see the importance of the factors specific to Asia in the same way and we were able to determine a typology of four groups of companies/respondents.

Let us look at each of these observations in a bit more detail.

Based on our initial set of 32 factors that appeared to influence how innovation management is practised in Asia, we were able to derive 10 consolidated factors that are perceived by our respondents to influence the management of innovation in Asia. In fact five of those explain more than 40 per cent of the variance. These five factors specific to Asia are:

1. The absence of an environment in which it is easy to operate as an innovator
2. A typical underdog mentality leading to a self-fulfilling prophecy of underperformance in innovation
3. A lack of good knowledge and human resources
4. A set of management approaches associated with traditional management in Asia that hinders innovation
5. The absence of appropriate management models for innovation management in Asia.

When we go back to Tables 3.4a and 3.4b, in which we gave the ranking of the most important hurdles, it should be clear to you that there is a consensus that the biggest difficulty is the inability to create a sustainable competitive advantage with innovation due to a lack of protection against imitation (not only because of IPR issues but also because there is quick imitation by competitors) as well as a lack of project management capabilities. The lack of resources may be an important influence on the innovative capabilities, but one that, on the average, counts less because it is a preoccupation of a smaller group of respondents (in particular the cluster of the 'poor in knowledge resources').

We did expect that managers from Asian and non-Asian companies would have a different view on management of innovation. It turns out that the differences are minimal, and, in general, managers from Asian companies are slightly more positive about the environment and about the capabilities of their staff.

If implementing innovation in Asia is indeed different, that does not mean these differences are similar for all respondents. In fact we derived four groups with different characteristics. Different attitudes towards the difficulties of managing innovation require different solutions. For management educators and policy makers it may be worthwhile to understand that the four groups require different types of support.

- The innovation starters do not blame the environment, but do not yet know how to innovate very well. This group probably needs more basic management education on innovation and project management.
- The tradition fighters feel the burden of traditional Asian heritage. They need to be able to escape this tradition and perhaps the infusion of new employees coming from different environments can help them.

- The poor in knowledge resources need access to engineers, designers, managers and knowledge. They may benefit from initiatives by the government or private organizations that enable technology and people transfer.
- The stuck in the muck think they can innovate but feel the environment is not right. They are the ones that may request changes in government policies that enable innovative behaviour.

Perhaps this is a moment for you to pause while reading this book. Do you recognize your opinions or your company in one of these groups? Perhaps you should go back to the ten consolidated factors and rate whether you think these are either high in importance to you as a hurdle for implementing innovation or not really important. And perhaps you should also go back to Table 3.5 and evaluate whether you see yourself or your company close to one of these groups. Doing this exercise will focus your mind in reading the rest of this book.

CONCLUSION

We started with the question of whether innovation management is different in Asia. Our two-phased research study led us to conclude that the principles of innovation management in Asia are the same as elsewhere, but that the implementation of these principles may differ. This is in line with some previous research.[10]

The results of our survey of 336 senior managers indicate that innovation is becoming more important and the perceived hurdles to implemention enable us classify the respondents in four clusters or groups: the innovation starters, the tradition fighters, the resource poor and the stuck in the muck. Belonging to these four groups is associated with the perceived competitiveness in the environment. The solutions provided by managers, management educators and policy makers to help Asian companies to become more innovative may have to be adapted to the different characteristics of each of these groups. You may actually try to identify with which group or combination of groups you feel the highest affinity. It will focus your mind when you read the next chapters.

In Chapters 4–8 we provide a number of examples that we hope may inspire you to overcome the hurdles for implementing innovation in Asia. Each of the chapters responds to a combination of some of the hurdles described above. In Chapter 4 we tackle the issue of renewal and unlearning. This is a response to several of the hurdles: the lack of an environment in which it is easy to operate as an innovator, the lack of a perceived need to innovate, the inertia created by traditional Asian management styles and the negative impact of a traditional cost reduction attitude on innovation. In Chapter 5, where we address markets and marketing, we show you how to overcome the lack of good market understanding. In Chapter 6 we describe how to gain access to knowledge resources. In Chapter 7 we focus on the returns that need to come from innovation. This chapter addresses two hurdles: the lack of external business rewards for innovation and the absence of some basic management models specifically applicable to innovation management in Asia. In Chapter 8 we show how to overcome the negative impact of an underdog mentality.

Since the challenges for each of the four groups are perceived to be somewhat different, you may want to read each of the following chapters from a different angle. The poor in knowledge resources may gain more inspiration from Chapter 6. The tradition fighters will get more out of Chapter 4. The innovation starters will perhaps be more interested in Chapter 7. And the stuck in the muck group will learn a lot from Chapter 8. But then again all the examples we provide may give you new insights.

Notes

1 See Appendix for the list of companies we studied through desk research or interviews.
2 De Meyer A. and Garg S., 2004, What Makes the Implementation of Innovation Management in Asia Different?, INSEAD Working Paper no. 2004/81/ABCM/TM. De Meyer A. and Garg S., 2004, Innovation management in Asia: some preliminary findings, in Albe S. (ed.) *Cross Functional Innovation Management: Perspectives From Different Disciplines*, Gabler, Wiesbaden, Germany.
3 De Meyer A., Williamson P. and Chua C.H., 2003, Banyan Tree Hotels and Resorts: Building an international brand from an Asian base, INSEAD case study, Singapore.
4 Kim W.C. and Mauborgne R., 1997, Value Innovation: The Strategic Logic of High Growth, *Harvard Business Review*, **78**(1): 102–13.
5 De Meyer A. and Chua C.H., 2000, MyWeb.com: Bringing the internet to the living room (A1), INSEAD case study, Singapore.
6 De Meyer A. and Bhardwaj S., 2003, Biocon, INSEAD case study, Fontainebleau.
7 De Meyer A. and Bhardwaj S., 2003, Pinoy2Pinoy, INSEAD case study, Fontainebleau.
8 Ettenson R. and Klein J., 2002, Branded By The Past, *Harvard Business Review*, **78**(6): 28.
9 See 'The Top 100 Brands' by InterBrand (www.interbrand.com).
10 Zain M.M., Richardson S. and Khan A.M.N., 2002, The Implementation of Innovation by a Multinational Operating in Two Different Environments: A Comparative Study, *Creativity and Innovation Management*, **11**(2): 98–106.

4 Creating new organizations in Asia for the new challenges

We learned from our case studies and survey research that many managers still complain that the traditional organizational structures in Asian firms hinder innovation. One of the first priorities for the firms that want to innovate is to explore how they can adapt their organization and organizational culture to make it more conducive to innovation. If you feel you can identify with a tradition fighter, as described in the previous chapter, the ideas that we develop here will be of high priority to you.

Since the rise of Japanese firms in the 1970s, the way Asian firms operate has attracted a fair bit of attention from management scholars based in Western Europe and the US. The attention shifted from Japan to East Asia in the 1990s when some of the Southeast Asian tiger economies grew rapidly and in a manner apparently different from that of Western firms. Over time, it has generally been accepted by non-Asian firms and observers that due to different social, economic and political systems, managing firms in Asia is a somewhat different ball game.[1]

We believe this still holds true. But the traditional approaches to management in Asia have not been adapted enough to stimulate innovation. It is clear that firms in Asia need to change the way they have been managing themselves. This is primarily because of the increasing opening and integration of Asian economies in the world trading system. Past government policies that shielded local firms from international competition are changing due to domestic and international pressure. The pace varies across countries but the direction is clearly towards a more open trade environment. As international firms enter or take deeper roots in the home turf of Asian firms, Asian firms need to respond in a multifaceted way, both in home and international markets, to be effective competitors. As we discussed in Chapter 1, innovation is one of the most important elements in the competitive strategy for many firms in Asia. We discuss many of the issues related to innovation in this book. The first one, however, is to put their house in order, that is, to make their internal organization right for innovation.

A new world requires new methods for success. First, Asian firms need to *unlearn* their sources of competitive advantage that served them well in the past. At the same time, they also require a considerable mental shift to *learn* the new sources of competitive advantage. As much of the growth in the future will be driven by innovation, Asian firms need to create an environment that fosters *creativity*. In this chapter we discuss these challenges and present some of the companies in Asia that have been able to transform themselves for the new world.

UNLEARNING PAST TRUTHS

Change management always starts with unlearning past truths, before one can put into practice some of the new ideas. In the next sections we mention two of the past mantras of Asian firms and show how they have become less appropriate or even counterproductive to the development of innovation.

Mantra 1: Privileged information and physical assets are the key to competitive success

Since the late 1980s, different parts of Asia from India to South Korea have undergone economic transformation. Irrespective of which political group governs in any of these countries, there has been a greater emphasis on liberalization, deregulation and integration with the world economy. The Asian financial crisis of 1997–8 was a pause in this trend in certain countries, for example Malaysia and Indonesia. However, as had been planned and since the crisis has been stemmed, these countries have become more integrated in the world economy and are in the process of developing a better regulatory framework.

Such trends have significant implications for firms in Asia. During the early stages of development, Asian firms first relied on a significant asymmetry in access to information. Later in their development, they relied on the physical infrastructure they had managed to build due to the privileged access to various financial subsidies and special permissions from the government. This government support is fading away more rapidly than anybody had imagined. In fact, in many countries in East Asia, the corporations are unhappy with the governments for breaking this tacit social contract of support.

Increasing transparency of processes, technology-enabled access to information and the relative ease with which physical resources can now be accessed or replicated have made the past benefits much less of an advantage. It is, therefore, important to move from relying on past advantages to thinking about knowledge as the critical resource that will give competitive advantage both in home and international markets.

The difference between knowledge and information is that knowledge is created by making sense of the information. The transformation of information into knowledge often requires a certain belief about its truth and what makes it actionable. It is often tacit and cannot easily be codified. We are not suggesting that privileged access to local information and control over physical assets are not useful. In fact, when these things can be combined with a knowledge focus, the company will certainly be in stronger position because of a better business system.

Let us take an example at the level of a nation. Until the mid-1990s, Singapore's national competitive strategy had largely been based on an important physical resource: its location. Singapore prospered through the petrochemical industry, the port, and high-tech manufacturing. All these relied on location to a large degree, supported by its superior physical infrastructure, efficient workforce and pro-business government policies (mainly financial incentives for foreign investors). Now, Singapore's competitive position is under attack, as the countries in the rest of Asia have a good and sometimes even newer infrastructure, technology makes location less important, the workforce for many industries is cheaper in other countries and the governments of neighbouring countries undercut Singapore by providing more financial incentives. Having realized the importance of repositioning its economy as a knowledge-based economy, Singapore is encouraging a focus on R&D and knowledge creation by building the necessary infrastructure. Since the late 1990s, billions of dollars have been invested in revamping the education system, transforming the public library system, creating the physical infrastructure and a supporting ecosystem of policies that attract world-class companies and talent in IT and life sciences. The port authority has been trying to use the sophisticated physical infrastructure to offer a plethora of value-added services to its clients, not just by itself but also with partnerships around the world.

Li & Fung represents such a transformation at the firm level. Li & Fung started in Guangzhou, China in 1906 as a trading broker. Being able to communicate with both American merchants and Chinese factories, its commission was 15 per cent. It became a regional sourcing agent for the Pacific rim when the need for trading companies diminished. As a sourcing agent, it helped to find good sourcing deals for its clients in accordance with trade quota restrictions through its regional network. But, as the region became more attractive and started opening up, companies started setting up their regional offices and the need for Li & Fung's role diminished again. In the 1980s, it started to offer much more value: the customer just described the product concept and Li & Fung found the best deals in the region and manufactured the goods and delivered to the customers. When Hong Kong, its home base, became less competitive for manufacturing in the 1990s, it dissected its value chain and outsourced work to southern China which also offered many incentives. Having mastered the art of 'dispersed manufacturing' and with the aid of new technological infrastructure, today Li & Fung is one of the leading global supply chain managers of worldwide production programmes for its customers. The customer only needs to specify what it wants and where and Li & Fung argues that it can put everything into place. Li & Fung doesn't own any production facilities, but manages over 7500 quality-conscious and cost-effective suppliers in more than 40 countries all over the world. It has three distinct core businesses: export trading, distribution and retailing, but its core competence is global supply chain management – a knowledge business.[2]

Mantra 2: Low cost is *the* source of competitive advantage

In the past few decades, most East Asian companies did well due to their lower cost advantage (and thus prices) as compared to what could be achieved in Western Europe and North America. Thus, many companies believe that lower price is the key to competitiveness – although many of them refer to 'low cost', it is actually 'low price' they should be referring to. However, as production is dispersed globally and China has risen as the world's manufacturing centre, low price is not a possible positive differentiator for companies in the rest of Asia or even Chinese companies in the coastal provinces. Without expertise in managing global supply chains and with no established brands, many low-price suppliers are lowering their prices even further, but it is clear to many that it is a losing battle. An example of this is the textile industry, where the combination of low labour costs and relatively high productivity in China makes it an almost unbeatable competitor. Until 2004 producers from countries like Bangladesh or Sri Lanka could still compete because they were favoured by the quota system imposed by the USA and the EU. But with the disappearance of quotas, Chinese producers will become very dominant in textile production. Pre-empting the negative impact that such dominance may create, at the end of 2004 China implemented a self-imposed export tax on some textile products in order to reduce its competitive strength.

Cost reduction, provided the quality doesn't worsen, is good from an operational efficiency point of view because it means higher productivity. But operational efficiency is rarely a sufficient differentiator, while there is overcapacity in many industries around the world. Thus, instead of getting a bigger slice of the cake by lowering prices, Asian companies would be better off with a focus on creating more value.[3] Only higher value can command higher prices. Here are a few examples to illustrate this.

Higher value is created for a company when it offers what people want, at a price they are willing to pay . Haier of China offers white goods in the Chinese market that reflect a thorough understanding of customers and their behaviour patterns, commanding high prices and loyalty. In the US too, Haier has established significant market share and a brand name in products such as wine coolers and compact refrigerators, again reflecting an understanding of its customers. One of its attractive new products is a mini-refrigerator that can serve as a computer table – ideal for small student rooms. It is able to rapidly innovate for the US by designing in the US itself a large number of the products sold there. Unlike a vast majority of Chinese products in the US market, the Haier products are not positioned as low-end products competing on price but as high-value products offering design and value to the customer.

Similarly, iRiver of South Korea has successfully and intentionally positioned its MP3 players as lifestyle products with aesthetic design around the world. The company likes to compare its products with jewellery and high-end watches. An iRiver MP3 player retail price is 10–20 per cent higher than other popular brands, even though the company was one of the latecomers in the industry.

Dilmah Tea from Sri Lanka has taken an important step to not compete on providing low-priced raw material to foreign tea brands. Since 1988, it has had its own innovative products and brands that are considered premium and differentiated in markets internationally. Dilmah capitalized on the globally appreciated high quality of Ceylon tea. Highlighting the fact that international brands mix tea from various origins, so diluting the final taste, Dilmah introduced the single region tea, which stresses the comparison between tea and wine, in colour, structure and 'mouth feel'. Fancy T-Bars have been opened in Sri Lanka to attract the young generation, and are also planned for the international market. Today, the Dilmah brand is sold in 90 countries and competes with the very best in the tea industry. Its teas are conspicuously stamped 'Packed in Sri Lanka'.

Until the end of 1990s, Samsung Electronics was largely an OEM of semiconductors and computer monitors for the large (mainly Western) multinationals. It had begun to try to market its own branded consumer electronics goods but they were poorly received and Samsung was still seen as a low-end, low-price company. Stung by the Asian financial crisis, Samsung was forced to transform. It allocated a large marketing budget to find new opportunities.[4] The efforts led to vast improvements in resource allocation, geographies and product categories. Today, it is a recognized world leader in three product categories: consumer electronic products, semiconductors and LCD (liquid crystal display) screens. According to branding consulting firm Interbrand, it is also one of the fastest growing brands in the world and closing the gap with its more well-known rival in consumer products – Sony. The less successful performance of Nokia in 2004 was attributed to some extent to Samsung Electronics, which is now accepted by Nokia as one of the main competitors in mobile phones.

Why is it difficult to unlearn?

We believe that it has been difficult for Asian firms to unlearn their reliance on information asymmetry, control over physical infrastructure and low-price strategies in the competitive game because of two important factors. First, companies find it easier to change when they have a role model to follow.

Most management books come from Western business school academia and offer examples that are often irrelevant to Asian companies. They are so much out of context that the companies sometimes find it difficult to relate well to the ideas explained in them. And those companies that try to import the ideas as touted do not always see the results.

Second, the remaining government intervention in Asia slows down companies from developing competencies required for global competition. The negative examples of interventions include the diversion of resources through development banks, requirement of licences and other permits to carry out business activities and offers of incentives for exports. While many of the incentives were created for developmental purposes, they became counterproductive when the firms started to overrely on them and exploit the system. Many of the largest firms across Asia, in some form or another, have until recently benefited from many of these government programmes, failing to create in their organizations a true sense of competitive spirit in the global marketplace.

MAKING A MENTAL SHIFT

The second step in change management is to develop a new approach or a new concept for your organization. Having understood the mantras that keep you from being an innovator, you now have to get into action. Based on our research, we offer you three main ideas: the stimulation of true value-creating entrepreneurship, the implementation of careful process management and the change from a product mindset to a capabilities mindset. We will also offer you some inspiring examples of how firms in Asia have implemented these ideas.

Entrepreneurship as risk management

As the global business environment becomes more uncertain and complex, sustainable growth will only be achieved by developing a culture of corporate entrepreneurship. Entrepreneurship is essentially about value creation and careful risk management. The lack of a risk-taking culture is often the reason behind limited corporate entrepreneurship in Asia.

Whenever we presented this statement in East Asia, somebody in the audience would invariably point out that the ethnic Chinese have a risk-taking culture and bet heavily in casinos and on horse races. But that is exactly our point – the risk taking implied by entrepreneurship as we described it in Chapter 2 is different from risk taking in gambling. In that sense, Asian companies are often not entrepreneurial enough to create new value through products and services. They lack a grand vision and a planned approach to achieve it while managing the risks properly.

It is of utmost importance for Asian companies to start thinking of entrepreneurship in terms of proper risk management for commercial gains. And such gains are sustainable only when rooted in knowledge and not privileged information as in trading. As discussed earlier, Li & Fung has become exactly that over time: from a simple trader to regional supply chain manager and now a leading, global supply chain manager. It is not about trading any more at Li & Fung. The whole business is created on a thorough understanding of industries and the ability to manage all kinds of risks in a global supply chain. That is the kind of risk management we talk about in entrepreneurship.

There is another challenge for entrepreneurs in Asia. Social stigma is often associated with failure in Asia. Corporations must create ways to minimize that effect. One way is to have a prototyping culture where the damage in cases of failure is limited. This is what is practised with every single innovation at the National Library Board of Singapore (NLB).[5] NLB manages Singapore's national library. Until 1995 it had the traditional characteristics of any national library, that is, a somewhat stodgy deposit of national and international books. Since the mid-1990s it has embarked on successive waves of innovation in order to bring media (and not only books) closer to Singaporeans, and develop a strong sense of customer service. The creative ideas developed during this transition were evaluated according to certain guidelines, and a prototype was created at a given library for every innovative idea. After careful observation, the prototype was evaluated to see if it should be scaled up. If yes, the learning of the prototype experience was incorporated and the innovation was rapidly scaled up and established across the public library system of Singapore.

From improvisation to careful process management

Firms in Asia often have a relatively superficial engineering mindset when it comes to innovation, that is, they love to make their product or service work without due attention to details and perfection. Improvisation is considered to be a good use of resources as it makes companies in the poor countries of Southeast Asia feel like winners. That is fine, but successful innovation is 98 per cent execution and only 2 per cent inspiration. Japanese manufacturing companies knew this well. They had enormous command over their processes, which made them successful at introducing new products and services to the markets. The rest of Asia needs to realize that improvisation is good, but reliance on improvisation alone without developing processes over time is suicidal in today's competitive environment, where imperfect products will only be tolerated in the market for a short period of time.

The NLB of Singapore realized this, and has achieved the reputation of being an innovative organization. It underwent a major business process re-engineering exercise so as to improve all its internal and customer-facing work processes, resulting in a massive improvement in performance and customer satisfaction. Another advantage of the exercise was that the power shifted from individuals to processes. It was facilitated by the use of IT, led to transparency and enabled continuity of services. The NLB also trained all its employees in project management methodologies. Such training not only improved the quality of execution, but also provided everybody with a common language, especially for cross-functional teams that are so central to innovation.

Processes bring quality and predictability, both of which are important. Let's take a more in-depth look at a less well-known company: Aapico of Thailand. Aapico Hitech plc is a Thai auto-parts manufacturer which made the headlines in 2002 when it won the contract, in the face of stiff competition, to design, test and supply jigs – a device to hold car parts steady as they are welded – for the Mercedes-Benz E-class to be built locally in Thailand from 2003. Aapico supply the jigs 20 per cent cheaper than DaimlerChrysler's usual supplier, making the contract a win–win situation for both parties. Aapico designed, manufactured and tested all the jigs for the new Mercedes-Benz E-class in Southeast Asia and other plants around the world. It was the first time that the entire jig for a Mercedes-Benz car had been made in Asia. And for Yeap Swee Chuan, the Malaysian engineer who founded Aapico 17 years earlier, it was a milestone towards his ultimate goal of transforming Aapico into a total system supplier encompassing all levels of the production process.

When Aapico started making low-volume jigs for these big car manufacturers, for example DaimlerChrysler, it had limited financial resources and little engineering or design talent. But, as Yeap Swee Chuan, its CEO, freely admitted, it has been the relentless focus on consistency and detail that has made Aapico successful with its clients. To operationalize such focus, Aapico developed internal processes. In fact, when there were opportunities to earn more money by taking on more projects, Aapico chose to focus on a few and not spread itself too thinly to make sure it didn't take short cuts in its processes. As a consequence it has developed some of the best concept designs for low-volume tooling in the world. Its better design processes have enabled Aapico to race ahead of its competitors.

Building Aapico into such a process-oriented, learning-based organization was no easy task. It required crucial investments in training the workers in new technologies and modern management techniques. This was where the CEO encountered a problem – recruiting qualified engineering graduates. It was not a question of quantity, but of quality. A string of flawed policies had ensured that most graduates in Thailand had not acquired the necessary skills that the private sector required. Undeterred by these shortcomings, Yeap took things into his own hands and recruited an experienced engineer from Ford in order to set up an in-house jig-design and engineering facility. Then in 1991, Aapico became the first among its peers to introduce CAD/CAM (computer-aided design and computer-aided manufacture) technology and CNC (computer numerical controlled) machines. By constantly challenging itself the company has managed to ramp up its technical expertise without overextending its human or financial resources.

Asian companies will need to learn how to manage the diffusion of know-ledge in their organizations. The complexity of this diffusion process increases as companies become more international. Asian companies will need to establish processes and tools for collaborative discussions to manage the old and new knowledge in a professional and purposeful way.

From products to capabilities

For the age of flexibility, it is better if firms consider themselves to be bundles of capabilities instead of products. The difference between these two approaches is that the different permutations or combinations of capabilities needed for existing product(s) can be used to create new products and services. So, once companies shift their mindset to capabilities, they can become a lot more flexible and empowered.

Tata Indica, India's first totally indigenous car, was conceived in the 1990s by the Tata conglomerate. The objective was to get into the business of consumer cars suited to Indian roads and acceptable to Indian purchasing power. The conglomerate, which had a successful truck manufacturing division, already had various technology pieces in place but dispersed across companies. Assembling those pieces of capabilities together, the company was successful in developing, producing and later significantly improving its Tata Indica. In 2002, Tata Indica was relabelled and sold as MG Rover in the UK and in 2004 was exported to the left-hand drive markets of Continental Europe as well.

The National Library Board of Singapore has taken the paradigm of capabilities to a different level. It has a business group whose objective is to package and sell the internal capabilities of the library to the corporate world. One of the most innovative things it has come up with is to use the capabilities of reference librarians, who have special but underutilized expertise. For corporations, the reference librarians now help to conduct brainstorming sessions. They then return to the corporations with research on some of the short-listed ideas, providing a basis for solid discussion for managers in charge in the corporations.

All these capabilities do not have to exist internally. They could be learned from partners and then combined with the existing ones. For instance, Patkol of Thailand learned about the problems that its various clients faced when they built new plants. Patkol is well known in ASEAN as a leading company in refrigeration and food machinery. It provides a range of ice-maker machines, cold room and freezing systems. Patkol also develops specific and turnkey projects for dairy, soft drink and fruit juice production plants, distilleries and breweries. It provides tailor-made food processing lines, manufactured to the requirements of each of its customers.

As Patkol was involved from the beginning to the end, when installing its special tanks, it got to know first hand the fundamental problems of its customers. More importantly, it also saw that the commercial returns were good; skills and time available with the company to develop the project were limited. Having learned the basics, Patkol offered the projects on a turnkey basis to less demanding customers to further refine its capabilities. Now, it has a fast growing turnkey business.

STIMULATING CREATIVITY

Having unlearned the past and developed some ideas of what the new organization requires, we also need to learn how to exploit the possibilities offered by the new organization. That requires creativity. Asian culture is often blamed for not promoting creativity. While that may be an exaggeration, there is probably room for improving the environment and making it more stimulating for creativity. What can be done? Based on our research, we propose three ideas: promote diversity among the employees, increase employee motivation and stretch the minds of your employees. We also offer you some inspiring examples of Asian firms that have implemented these ideas.

Diversity of employees and communication among them

Asian companies tend to have limited diversity among key employees, be it in terms of nationalities or educational and career backgrounds. If one agrees with the widely accepted notion that diversity encourages creativity, Asian companies certainly underperform. Sometimes, experts are brought in to fill the knowledge gaps. For example, Smart Communications, a leading mobile phone provider in the Philippines, hires some of the best experts in the area of mobile telephony security. But the focus as such is not to induce creativity in the organization.

As the Indian IT industry targets customers around the world, Indian IT companies are trying to make their workforce more international by hiring more people locally in different countries. The Indian company Infosys Technologies is a NASDAQ-listed company that provides consulting and IT services globally. It describes itself as a partner to conceptualize and realize technology-driven business transformation initiatives. With over 32,000 employees worldwide, it is recognized to have developed a low-risk 'global delivery model' to accelerate IT project schedules with a good degree of time and cost predictability. Infosys, which derives a large part of its revenues from clients outside India, realizes that being multicultural gives an important advantage, especially when bidding for international, high-value contracts.[6] For instance, on large deals it ensures that people from different parts of the world contribute, on a collaborative basis, to prepare, defend and to execute a proposal. Its 32,000 employees come from 38 different nationalities but are still predominantly Indian. The Indian IT industry is realizing that managing diversity is a challenge. For instance, it has not been easy for Infosys to transfer employees from the US to its headquarters in India, although moving people from the US to Europe or across various functions has been easy. Getting managers to have mutual respect and trust and then collaborate is also a challenge. At the Infosys Leadership Institute, as at the leading management development centres such as that of GE in the USA, Infosys senior executives teach themselves courses and help to integrate Infosys managers across the world.

Employees of some of the Singapore government research labs are encouraged to go and work with SMEs for a period of time during their employment at the labs. It is believed that such an experience will create different perspectives and help the research labs to be more creative, entrepreneurial and relevant in their research when the employees return.

Similarly, playing with the organizational structure and forcing employees to work together in cross-functional teams often unleashes creativity. Not only does that happen at lower management levels at the NLB of Singapore, but it is also true at the highest level. One of the main reasons behind the vast efficiency improvements and many innovative services at the NLB has been the composition of the senior management team. For the first time in 1995, the senior management team (including the CEO) comprised not only librarians but also businesspeople from the commercial world, which led to a radical rethink of the library system and new creative ideas.[7]

Easy communication within an organization is the first step towards a creative environment. South Korean companies are not known for lateral network communication in the organization. The communication is mainly top down. These communication styles are deeply rooted in the societal system of South Korea. Neowiz, a South Korean company created in 1997, offers internet services, in particular in online gaming, online sales of avatars and web-based community services for example, chat groups. To improve the quality of communication and respond to the rapidly changing competitive environment, it has developed a 'cc culture' (cc refers to carbon copy), essentially a comprehensive email sharing culture at all levels of the hierarchy and across the organization. For instance, when a product is launched, 60–70 comments are generated on emails from employees at all levels across the company. As people receive information, they are encouraged to participate in brainstorming or giving feedback. With such a high level of sharing emails, people are also forced to take full responsibility for what they commit to. It takes new employees a while to get used to the level of transparency and participation, but soon they become active participants. The challenge will be how to continue the system and manage the number of comments when the company grows from a small workforce to thousands of employees.

Employee motivation

People outside Asia often have the image of people working long hours in Asian companies. While this is true, being at work is different from really working with full heart and mind. An efficient workforce is a different thing from a creative workforce. This is a challenge in Asian companies. The 'hardworking' workforce keeps the bosses happy, but whether that workforce is really efficient or creative is not so obvious. For instance, a 2003 study by Gallup found the workforce in Singapore to be the most disengaged workforce in the world. There are two reasons for this: a family-centred senior management team and an unsuitable incentive structure.

Don't misunderstand us. There is nothing fundamentally wrong with the family-owned businesses. Around the world many of them are seen to be highly effective organizations. A study reported by *Newsweek* in 2004 found that family-controlled companies outperformed their rivals on all six major stock indexes in Europe.[8] And some family-owned companies are among the most innovative in the world. However, there is a difference between the Asian family-controlled firms and the North American or European family-controlled firms. Due to the lack of a proper, efficient legal system in most of Asia, families still tend to control the management of their companies to a large degree by appointing their own family members, close friends or associates who would carry out what the family wants. This demotivates other employees as they know that there is a 'glass ceiling' for them. The loss for the company is that it is only able to tap into a small amount of the available talent. Family control also creates a bottleneck for rapid decision making and stifles entrepreneurial behaviour at lower levels in the organization.

Samsung Electronics, the flagship company of arguably the strongest chaebol in South Korea, has made conscious efforts to change the family influence in management. For instance, the CEO Yun Jong Yong is not a family member and there are hardly any family members now in the senior management team apart from the chairman. After the Asian crisis, the company cut a third of its workforce and replaced half of the senior management with younger university graduates. The company now gives its employees a degree of autonomy rare in South Korea and in Asia overall, linking pay to their contribution to profits. The company has also appointed foreign board members. The corporate culture aims to defy South Korea's tradition of lifetime employment and blind respect for authority. And all this is because of the competitive pressures of increasing globalization and a need to unleash the contribution from employees.[9]

Similarly, Reliance Infocomm, a new information and telecommunications arm of Reliance Industries in India, is fundamentally different from the rest of the Reliance group. Most of the Reliance group, which operates in traditional and commodity industries, is run in a top-down manner. However, Reliance Infocomm, which operates in a highly competitive industry, is organized so as to empower the people at all levels of the organization. This is one of the main reasons why Reliance captured the top position in less than a year of its operations, despite being one of the last entrants into the industry in India.

Incentive structures are critical in stimulating creativity and getting the best out of the employees. Many Asian corporations today are unable to effectively tap into the creative energies of their employees and other partners because they are unable to create a proper incentive system, financial or otherwise. The knowledge-based industries of IT and pharmaceuticals in India, South Korea and elsewhere in Asia are exceptions because they manage their incentive structures, where possible, close to global standards.

The lack of a proper incentive system has led to a traditionally high employee turnover in Asia. This is why companies do not like to share knowledge with their employees and do not encourage discussion among them because they fear that employees would leave to start a competing firm. Patkol of Thailand, on the other hand, uses another strategy of sharing information with its employees and encourages discussion among them. To try to keep employees in the company, Patkol has created unprecedented incentive systems to attract and retain bright people. These include a family-like yet professional work culture and stock options. Despite being a small company, it recognizes the value of motivation and its link to performance.

At Neowiz in South Korea, one cannot be considered for a promotion if one has not proposed and tried new ideas. The company believes that if one has never tried, one will not fail. Lack of failure doesn't mean much and thus employees are motivated to try their best.

At Haier of China, the performance responsibility is distributed across the department. Each employee in a given department has three financial statements: balance sheet, profit/loss, and cash flow statement.[10] Even R&D employees have a market objective, and they are judged not on the basis of what they design but how well the designed products sell. Such an incentive structure forces R&D employees to take the initiative, talk to customers and be creative in addressing their *needs*.

South Korean companies are not known for giving a free hand to their managers, especially foreign managers. However, Lolita Lempicka, a leading perfume line based in France but owned by the South Korean company AmorePacific, is run by European management with significant autonomy. Such autonomy is required for two reasons: the perfume industry is a creative one and people do not like to be dictated to; and the Korean owners have little to contribute in terms of geographical and product expertise. Thus, autonomy is the best way to achieve performance from its French employees. More will be said about this company in Chapter 5.

Stretch your employees

A good working environment that encourages creativity is important, as is employee motivation to tap into such an environment to do something creative. However, the presence of some pressure not only ensures that employees continue to think creatively, but also helps to direct the creative efforts towards a particular larger goal. How do you create this pressure?

Capital markets seem to be a prime candidate in creating pressure on companies. However, debt financing instead of the more pressurizing equity financing is still more popular in most of Asia. India is the exception where the equity culture is deeply rooted. In the case of debt financing, creditors seldom care about how companies are investing in growth as long as they get their interest and the principal back. In fact, some of them take undue risks because they know that some companies will not be allowed to fail by their governments and their returns will be guaranteed, irrespective of the debtor's performance. The family's own investment in family-controlled companies is really patient capital and the prime objective is to keep control within the family.

Equity investors, local or foreign, tend to sell their holdings rather than challenge management on its behaviour.[11] Even though increasingly since the 1997/8 crisis a large percentage of stocks are owned by foreign investors, governments still tend to help local companies, especially prominent ones, during shareholder uprisings. Thus companies often have to look for other means to put pressure on their organizations. Changes in governance can be one of them.

Samsung Electronics has taken many voluntary steps to create pressure and independent monitoring for itself that it thinks it needs in order to be an innovative global player in consumer digital electronics. For instance, it has made major changes to the composition and function of the board of directors to strengthen its independence. In 2004, seven out of thirteen directors were from outside the company and three of the directors were non-Korean. The auditing committee is also made up entirely of outside directors. The company evaluation metrics are no longer only total sales and market share. An investors' relations committee is now dedicated to further enhance operational transparency. The effect of all this is that Samsung Electronics has made itself stronger by making itself more open and transparent to the world. Since all kinds of external forces require Samsung to perform, its management is always able to make some of these forces felt throughout the organization and hence channel the creative energies of its employees to reach the highest level of performance.

Exposure to best practice and ambitious targets can create another form of pressure. Aapico of Thailand was a small privately owned company until 2003. Being in a highly competitive global industry, the CEO had successfully communicated the idea that the pressure should and would be high for anybody who worked in the company. As he had not sold the company at the behest of its employees at one of the company's crisis moments, he always reminded employees that the company had a responsibility to perform. But how? The financial resources were limited and staff had average qualifications. He sent employees around the world and brought in visitors from around the world. That exposure not only created a first-hand sense of how competitive the industry was, but also a camaraderie as Aapico was investing to create capabilities. When the CEO returned from various seminars, where senior executives of top car companies had spoken, his message about the need for higher competitiveness was better appreciated. As the company achieved further success, the CEO has created a sense of pressure in the company by saying that Aapico could create a production system that would be better than the Toyota production system.

The National Library Board management created a sense of pressure for creativity in other ways. It referred to the lofty targets in the recommendations in the Library 2000 Report by a high-level committee comprising civil servants, academics, corporate executives and journalists. To create a sense of pressure, the management team pointed to the goals set for the Board. Easing the employment tension among employees, management promised that nobody would be fired due to the change. Management also highlighted that the pace and kind of work should change and everybody would be offered retraining if they wanted it, indicating the high expectations from management. Despite the fact that salaries were quite low at the NLB, the self-selected employees who stayed at the NLB were creative and hard working.

How much pressure should the company leadership exert to stimulate creativity? It is a difficult question and relies on intuition that is developed with experience. Hence, leadership is of primary importance in creating that delicate balance. The balance depends on a deep understanding of what makes employees tick. Understanding the culture at three levels – national, industry and company – is important. For instance, when the NLB pressured its employees, it used the national cultural trait of Singapore of not letting the government down on its objectives. Aapico, on the other hand, relied on the heavy industry culture of low tolerance of faults and high demand on quality and later the company culture of excellence to create a model Aapico production system.

Research on creativity agrees that pressure and stretch make people more creative, on condition that they can see the direction they have to work in. The above examples illustrate this – there is a strong stretch with a clear strategic direction.

CONCLUSION

In order to become a successful innovator, you may want to start by reflecting on how you can adapt your organization. This chapter has shown that you have to instil two beliefs that are different from the traditional management wisdom in Asia:

- Future competitiveness will come from emphasizing that knowledge, that is actionable information that we believe to be true, is the core resource of the organization
- Competing on low prices is a difficult battle; only higher value will command higher prices.

Some Asian organizations and firms have already made this mental step and we can be inspired by some of the actions they took to implement the new organization. Their organizations have made a mental shift and have invested in the creation of an environment that stimulates creativity.

The mental shift is helped by at least three levers. Company leaders need to make it clear that innovation needs a focus on value creation through entrepreneurial risk taking. Secondly, you must be able to make the shift from creative improvisation to careful process management in design and development. And thirdly, you have to see your firm as a bundle of capabilities that can be constantly recombined in order to respond better to the market, not as a portfolio of products.

Creating an environment that enhances creative behaviour also rests on the use of three levers. The first one is the creation of more cultural and cross-functional diversity within the management team. For such diversity to have an impact, you must ensure that everything is done to create good communication. The second lever is to increase the motivation of your employees through a more professional approach of your management, incentives and the decentralization of initiative throughout the organization. The third lever is to apply an appropriate dose of stretch: enough to render people creative, not too much to stop all their initiative. Capital markets and exposure to best practice can help you to realize this stretch.

Notes

1   Lasserre P. and Schutte H., 1999, *Strategies for Asia Pacific: Beyond the Crisis*, Macmillan – now Palgrave Macmillan, Basingstoke.
2   More details about this can be found in De Meyer A., Mar P., Richter F.J. and Williamson P., 2005, *Tigers Leap*, Palgrave Macmillan, Basingstoke.
3   There is some work done by Kim W.C. and Mauborgne R., 2005, *Blue Ocean Strategy*, Harvard Business School Press, Boston.
4   Corstjens M. and Merrihue J., 2003, Optimal Marketing, *Harvard Business Review*, October pp.114–22.

5　De Meyer A. and Garg S., 2004, The National Library Board of Singapore, INSEAD case study.

6　Gautam K. and Sinha J., 2003, The Next Hurdle for Indian IT, *The McKinsey Quarterly*, (4).

7　Garg S. and De Meyer A.,2004, op. cit.

8　See April 12, 2004 issue of *Newsweek International*.

9　De Meyer A., Mar P., Richter F.J. and Williamson P., 2005, op. cit.

10　Wu Y., 2003, China's Refrigerator Magnate, *The McKinsey Quarterly*, (3).

11　Barton D., Coombes P. and Wong S.C.Y., 2004, Asia's Governance Challenge, *The McKinsey Quarterly,* (2).

5 Markets and marketing

Innovation needs a different mindset, as we discussed in Chapter 4. But more importantly innovation begins with a customer. Without a comprehensive understanding of customers, innovation by a company is essentially its own private imagination. And, more often than not, such imagination is not commercially fertile. As we discussed in Chapter 3, a substantial disadvantage for innovation by Asian companies is that they are far removed from the sophisticated and leading markets in industrialized countries. Japan, the US and some of the leading European markets, are geographically far away and culturally different from Asia. Asian markets are perceived to be small and unreceptive to innovative goods and services. Even if attempts are made by Asian companies to offer innovative goods and services in such markets, customer irresponsiveness is a significant hurdle. In the survey described in Chapter 3, we saw that the lack of reliable market data and sophisticated customers score in the top six hurdles for innovation for the whole group of respondents. Lack of customer input during the development phase, the inadequate number of early adopters and the absence of customer feedback affect the quality of innovation. There is also a lack of reliable market data in most markets in Asia. Finally, the heterogeneity of Asian markets is a formidable challenge.

If innovation begins with customers, Asian companies clearly have some significant challenges. Many of them, therefore, choose the easier and less lucrative path of continuing to be non-innovative. Some take the alternative path. In this chapter, we discuss the challenges we introduced above in more detail and illustrate how some companies have *successfully* tackled them.

ASIAN COMPANIES HAVE TO DEVELOP STRATEGIES TO OVERCOME THE DISTANCE TO SOPHISTICATED MARKETS

A lack of pressure from markets makes most companies sloppy and complacent. The sophisticated markets in industrialized countries expose companies to customers who are extremely demanding and well versed in the product features and relevant technologies. Companies that want to thrive have to meet the demands of the market, which often forces them to put their creative energies together to address the demands. In other words, sophisticated markets have a role in inspiring if not forcing innovation.

Most Asian companies, however, do not go to developed markets for that reason. They go to those markets with the predominant objective of exploiting price arbitrage by exporting the products they make (and sell) in Asia. However, strategies like these often do not work and are unable to provide sustainable growth opportunities for companies. In many cases they reinforce the perception that Asian goods are low-end goods. Customers in sophisticated markets are much less price sensitive than in the Asian countries. They are willing to pay for quality and innovation. But, to know what customers in sophisticated markets want, Asian companies need to invest resources to profit from the opportunities. It is imperative for innovation that companies understand their customers very well.

Although we talk about the 'global village', real innovation is often not possible without being located close to the customer. Geographical proximity is almost essential to understand the non-codified knowledge about the customers. Codified knowledge is of course easy to obtain despite distance, but very often it is the non-codified knowledge that is the source of competitive advantage for companies today. For example, Kamthorn Kamthornthip, director of ChoLam, a Thai company producing ready-made clothing, told us in response to the question about how he could compete with Thai designs in Europe:

> In Thailand, we do not develop good fabric. One cannot use very thick and heavy fabric suited for very cold countries. In Thailand, we do not have the sense of cold climate ... We do not know how to design winter clothes.

This simple example shows that being physically present in sophisticated markets to obtain non-codified knowledge is important for Asian companies. Unless Asian companies can gain access to this intangible, non-codified knowledge, they will remain no-namers, poor brands and subcontractors. Or they may remain branded or unbranded suppliers to certain Asian ethnic groups, through outlets, for instance, in the Chinatowns that are present in many towns and cities in industrialized countries.

The journey for change begins in the mind. Asian companies must realize that without tapping these sophisticated markets for knowledge, successful innovation is unlikely. In other words, and to paraphrase a standard expression, 'when in France, act like the French' – as illustrated by the AmorePacific example. AmorePacific is a large cosmetics producer in South Korea. The company started its overseas expansion by exporting cheap make-up to women in developing countries. And it did a reasonably good job there. Then, in the early 1990s, AmorePacific tried breaking into the French market by exporting its skincare products. But they were received poorly. The Koreans couldn't understand the French preferences and lacked brand power.

The company's executives felt that the company 'ran into Gallic chauvinism as the French turned up their noses at cosmetics originating from a newly industrialized country'.[1]

The company realized its early mistakes. The chairman of the company was determined to get into the French market – the world's largest cosmetics market. After several unsuccessful attempts to acquire small French cosmetics firms, the company decided to set up a new perfume facility in France just outside Paris. Fragrances are the cosmetic products to which consumers show the least loyalty. Therefore perfume was the product chosen to launch the cosmetics line. AmorePacific hoped in this way to increase the probability of its success. The former international marketing director of Parfums Christian Dior was brought in to lead the European operation. The company hired French fragrance experts to formulate the perfume, and a French artist to create a charming bottle which was conspicuously stamped 'Made in France'. The perfume was brought to the market by French marketing and salespeople.

In a complete reversal of the past attempts to push a Korean brand on the French market, it was decided that a purely French brand should be created. A famous female French designer – Lolita Lempicka – was signed under an exclusive deal to use her name on a new line of fragrances and other beauty products. Even today few customers know that the perfume 'Lolita Lempicka' has a Korean company behind it.

Since the launch in 1997 of its French fragrance, AmorePacific has achieved outstanding success in the most sophisticated perfume market in the world – France. Lolita Lempicka was recently the fifth most popular perfume in France, outselling for example, Chanel's Allure and Yves Saint Laurent's Opium, which retail at a similar price (50m eau de parfum spray retails for about €60 in France). For a company from Korea, a country which doesn't have a tradition of Western-style perfume, it is an impressive achievement by any standards. Today the fragrance is available in over 90 countries. About 40 per cent of sales come from outside France. Success in France has boosted AmorePacific's confidence to invest abroad and compete in the top league. Bolstered by Lolita Lempicka's rapid growth, AmorePacific is now increasing its manufacturing capacity fivefold, with new production lines in France.[2]

Haier of China has adopted the same strategy of going to more sophisticated markets in the US and Europe. This strategy is unlike that of other Chinese companies that usually target the easier markets of Southeast Asia. Since 1990, Haier has been investing in building its brand in the US market. Its stated objective of going to the US was to build a brand and not just earn foreign exchange by luring customers with low prices. While low prices are something that Haier was able to offer due to the way its operations run, it competed on its in-depth understanding of customer needs and the speed to market. On this basis it has built a market share of more than 50 per cent in compact refrigerators and wine coolers in the US.[3] The compact refrigerators, targeted at college students who live in studios or small apartments, are popular because they address customer needs – the refrigerator has two wooden flaps on the side that can be folded out to make a computer table and folded back when not needed. Haier goods for the US markets are designed and produced locally in its own facilities in the US even though this means higher costs. The objective is to serve the market with goods that address local needs. Its strategy is to expand from niche consumer markets to bigger markets by continuing to learn from the market and introducing innovative products.

It is clear that to tap sophisticated markets, many different strategies can be followed. The two examples of AmorePacific and Haier illustrate successful attempts to establish a wholly owned subsidiary overseas that can play the role of a listening outpost. Another strategy could be to try to learn through an *alliance* with a Western partner. This may look more attractive because the investment will be smaller than in the case of a wholly owned investment. In the case of an acquisition, the price may be the same, but the speed with which one can learn may be much faster. But, as you would expect, these approaches are not always an easy road to ride.

Some top players in Asia such as Singapore Telecom (ST) and Singapore Airlines (SIA) have made equity investments in Western companies to build alliances. For instance, ST had an equity stake in Belgacom of Belgium and SIA has one in Virgin Atlantic (the holding company for Virgin Atlantic Airways, Virgin Holidays, Virgin Sun, Virgin's cargo operation and Virgin Aviation Services) of the UK. The key reason for such alliances may be to provide better services to clients by sharing infrastructure – telecoms networks in the case of ST and code sharing, enhancement of their frequent flyer programmes and shared access to passenger lounges and airport facilities in the case of SIA. Such investments are often offloaded (as in the case of Belgacom and ST) because they are made in companies that offer the potential for good financial gains. There is perhaps the intention to learn, but this is never stated explicitly and neither is it clear if ST or SIA have learned about European markets through such alliances.

When the Malaysian national car maker Proton bought the British sports car maker Lotus, its stated intention was to tap the R&D capabilities of Lotus. Lotus was to benefit with opportunities to develop new export markets for its world-renowned sports cars, particularly in East Asia and, together with Proton, to take part in the World GT1 motor racing championship. Lotus, which was already losing money at the time of the acquisition, continued to lose money until 2003 and there is a general consensus that Proton has failed to learn much from this acquisition about being more competitive in international markets.

Giordano of Hong Kong had to terminate its joint venture with SB Warenhaus (REAL) of Germany. Giordano had intended to sell the Bluestar Exchange trademark through REAL's hypermarket networks. Although

Giordano claimed to remain determined to enter the European market, the management acknowledged that 'tastes and culture in European countries are very different to Asian countries'.[4]

Such examples of failure should not suggest that alliances or acquisitions cannot work. Alliances certainly can open up opportunities. However, in making such learning alliances, Asian companies also have to offer something in terms of competence. Learning alliances work better when there can be some know-how trading, even if in an informal way. More often than not, the main advantage that Asian companies can offer to companies from the industrialized world selling goods in sophisticated markets is cheap manufacturing or services. That often leads to subcontracting relationships with Asian companies rather than learning alliances as such. Such arrangements *may* still provide Asian companies with an opportunity to climb the technology ladder, but they do not give opportunities to learn about the *market*.

One example of a company that was able to create a more trustworthy front end with an acquisition is the Samtel Group of India. Samtel has two types of businesses in the more general category of CRTs (cathode ray tubes) – one for mass markets like TV and the other for professional requirements like medical, military and industrial applications. In the former, Samtel sells mostly in the domestic market and about 20 per cent overseas. For mass markets, Samtel's products are the most expensive parts of the TV and the TV makers are concerned about this purchase. When Samtel exports, it finds that overseas customers treat it like their own domestic customers. They have demands and expectations like any rational, responsive customer.

In the latter, Samtel sells mostly in the export market (where most of the market for such products lies). Most sales are to the EU and the US. The CRT is an important but not the most expensive component in the customer's products. The segments are professional, for example for medical and military applications, and tend to be more cautious and conservative. Also, some of them are regulated. Here, Samtel found an unspoken resistance to buying from an Asian country. In the words of a senior executive of Samtel, 'there are real and imagined issues about procuring critical materials from a third world country so far away'.

Samtel's response to this phenomenon was to buy a German company in trouble and retain it as a technical and marketing arm. The customer's requirements were sought from Germany, the technical designs were mostly created in Germany and the servicing of supply was done in Germany. While doing this, Samtel was transparent that the materials were manufactured in India. With the comfort and strengths of the German unit available to customers, Samtel was able to gain their confidence now that the supplies from India were as good as they were getting or expected from Germany.

ASIAN COMPANIES MAY ALSO FIND SOPHISTICATED NICHE MARKETS IN THEIR OWN BACK YARD

Europe, Japan and the US are not always the most sophisticated markets for all kinds of products or services. For instance, when it comes to mobile phones or online games, South Korea is one of the most sophisticated markets in the world. Manila is the SMS capital of the world (although mainland China may replace it in the future). Whereas globally the average number of messages sent per subscriber per month is about 35 (with the USA and France both averaging less than 20), the Philippines has an average of more than 190 per month.[5] Thus, Nokia has even set up a group to learn how people use SMS to help to develop more sophisticated SMS-based applications. The electronic road pricing system in Singapore is a world-class example of a well-functioning implementation of road pricing, attracting government officials from around the world. The hospitality of hotels, resorts and airlines in Asia is among the world's best. Singapore is the world's busiest trans-shipment port. Since there is little space in Singapore, the containers are stacked high and yet the processing is one of the fastest in the world. This is due to the advanced application of operations research and the sophisticated tools developed by the Port of Singapore Authority and its affiliates. What these examples show is that some markets in Asia can be leaders in the world and thus be a source of global innovation.

Working out how to use the lessons learned in these markets to offer products in richer and bigger markets overseas is definitely a promising way for Asian innovators to be successful, rather than trying to beat US or European companies on their own playing field. This requires Asian innovators to overcome a few hurdles:

- South and East Asian markets are often either too small or too poor for profitable innovation
- Asian customers are unresponsive
- The lack of reliable market data in Asia and the heterogeneity of Asian markets.

Some companies have found clever ways to overcome or circumvent these preconceived ideas. Let them once again be an inspiration to you.

Overcoming the first hurdle: 'Asian markets are too small and too poor for profitable innovation'

Many companies tend to think only about the premium segments of the market. The fact that the low-income segment of society in Asia has low purchasing power makes them ignore the fact that this is an enormous segment in Asia. Quantity may compensate for a lack of spending power.

When competing in the premium segment, most Asian companies end up losing money in price wars with foreign brands and other Asian companies unless they control some critical resource(s). This phenomenon is very evident in China – the beer industry is a well-documented example, where almost all foreign companies competed in the small premium category and neglected the large-volume, low-end segment.[6] And if companies offer something for low-income segments, the products/services tend to be no-frills versions of what is offered to customers in the premium segment. But if companies address the lower income segments with truly creative, tailored innovations, the returns can be enormous.[7]

Smart Communications of the Philippines did exactly that. It observed carefully the low-income segment of the Philippines and then tailored services to suit this segment, achieving market leadership despite being a late entrant in the mobile telecommunications industry. In May 2003, it introduced a service called e-Load, which is conceptually 'airtime in sachets' and was the first of its kind in the world. In the Asian market the sachet is popular with shampoo users because it suits the cash flow of the ordinary consumer. When Smart Communications realized that 300 pesos for prepaid cards was high for the low-income segment in the Philippines, it had the insight that it could sell these prepaid cards like shampoo. More than 95 per cent of mobile phone users in the Philippines use a prepaid service and that proportion may be even more in the low-income segment. The high price of a card slowed the use of Smart's services. The comparison with shampoo in sachets gave the company a simple but powerful idea.

The company not only lowered the minimum amount to P30 (pesos), but also provided 'over the air' loading. In other words, people could get phone credits over the air in the smallest 'sachets' of P30. All of a sudden, mobile phones became an instrument which could be used by the low-income segment of the market. At the same time, phone prices were coming down dramatically. A secondary market of second-hand phones was being created as advanced models with cameras and PDAs started appearing. This encouraged people around the world to upgrade their mobiles. Second-hand black and white models of Nokia sold at US$35–40 in the Philippines.

Smart has completely changed users' behaviour and the distribution pattern of phone credits. With over-the-air loading, Smart has moved from being just a shop in a mall. It has about half a million agents and retailers all over the Philippines, which is a vast geographic area comprising many islands. This has created huge economic benefits through sales commission and employment. Smart sends the credits over the air with a simple SMS to its distributors and dealers, who have a special SIM card issued by Smart.

To offer such a service, over the years Smart has built a sophisticated and highly reliable back-end infrastructure. But the real reason why Smart is able to offer e-Load is because of Smart Money, an innovation it created a few years earlier. For this it won the 'Most Innovative GSM Wireless Service

Award', a prestigious international award from the GSM Association in Europe. The Smart Money service allows mobile phone users to make financial transactions on their mobile phones. For instance, it enables customers to use their mobile phones to transfer cash from their bank accounts to a MasterCard cash card. The transfer could be made to one's own cash card or to somebody else's. It took some time to convince the banks that Smart Communications was not competing with them in any way. It was actually expanding the market by tapping into the micro-transactions which would have taken place with cash: 95 per cent of all transactions in the Philippines took place with cash. In 2003 14 major banks had joined the Smart system. There are seven million people using e-Load, with about 2.5 million transactions a day. e-Load is more than half of Smart's outgoing revenue. The Philippines, consequently, is perhaps the most adoptive market for mobile commerce in the world.

In December 2003, Smart launched another innovation, called the 'Pass and Load', building on e-Load. The service allows users to pass prepaid credits to one another over SMS. Irrespective of the amount passed around, Smart gains SMS fees when credits are passed by SMS. It increases the speed of credits. The minimum amount that can be passed is 10 pesos, which is equivalent to less than two minutes of phone call or 10 text messages. Understanding customers' behaviour, Smart has worked out attractive schemes for loading that benefits it well too. For instance, a P30 transfer can be used within five days, but a P10 transfer can only be used within 24 hours.

In 2004, Smart came up once again with another innovative service. This time the objective was to tap into the eight million Filipino workers who are spread around the world and send home several billion dollars in savings annually. Usually the remittance by overseas Filipinos is made through bank and money transfer companies. Smart offered a new service through its remittance partners in 17 countries to send money home much more cheaply and instantly. The remittance partners send an SMS to the Smart mobile phone recipient in the Philippines. Receipt of the message and money can be confirmed by the recipient instantly, as it is reflected in his/her Smart Money electronic wallet account that can be checked on the mobile phone itself. If the recipient has a Smart Money card, the cash can be withdrawn instantly. Those who do not have the card can pick up cash immediately at any of Smart's partners such as McDonald's or 7-Eleven shops anywhere in the Philippines. Smart gets 2.50 pesos per transaction.[8] The total cost for the sender is much lower than going through the normal banking system.

In 2004 Smart Communications was widely considered to be one of the most innovative GSM wireless service providers in the world. Interestingly, none of the services are based on expensive 3G technologies and many of them are targeted towards the low-income group in the Philippines, itself a low-income country.

That Asian domestic markets can be lucrative despite being poor can also be illustrated by Tiger motorcycles of Thailand. Within eight months of its launch in July 2002, Tiger motorcycles had gained more than 3 per cent of the national motorbike market, which is the fourth largest in the world by sales volume. Tiger's success was a result of the attention paid to the needs of Thai customers who want style and efficiency but at a low price because of their limited purchasing power. Tiger employed no outstanding technology but it innovated and made designs especially for the Thai consumer. Its CEO Piti Manomai-Phibul, who used to be the managing director of Kawasaki's joint venture in Thailand, argued that the Japanese brands had looked at Asia as one market. That gave Tiger Motors an opportunity to outshine them in a local market.[9] Selling more than 5000 units a month, in 2004 Tiger bikes had overtaken Kawasaki, which was one of the four Japanese brands that had traditionally dominated the Thai market.

Japanese makers have cut the prices of their motorcycles up to four times, while adding unprecedented numbers of cheap models. Mass-market motorcycles in Thailand now cost less than Bt30,000. Pre-Tiger, they were around Bt40,000. It has been estimated that the entry of Tiger has cost Japan's four major players about Bt10 billion a year.[10] Nonetheless, Tiger's sales were rising and the customer response was good.

Tiger's competitive edge went beyond price. The company was not simply copying its Japanese rivals' models. Making its own parts in a Thai factory and employing local engineers and managers, Tiger saved money that was redirected to R&D. Tiger was planning to introduce three new models adapted to Thai market needs every year.

Along with original research and development for local market motorcycles, Tiger has developed toy Nano electric-powered minibikes for children. Similarly, there are special models for teenagers. Tiger also introduced a new model of its Nano electric motorbike in conjunction with the new Warner Brothers' movie *Torque*. Thus, focusing on large underserved segments of Thailand, Tiger has achieved commendable success, indicating that the Asian domestic markets can be poor but lucrative. Understanding customer needs is the key to success.

One often cited reason for not focusing on the low end of the market is the lack of distribution infrastructure. However, the distribution infrastructure is quickly becoming better and more accessible as government policies change. In any case, there are creative ways to get around the problem of rural distribution. Hindustan Lever Limited (HLL), the fully localized arm of Unilever in India, has found a way to get into the smallest villages. To compete in the low end of the market, controlling cost is very important. So, it sourced support from the rural areas of India, and offered rural people the chance to be entrepreneurs by offering distribution opportunities and training them. Not only do these village people know their surroundings better than HLL, but they are also grateful to HLL for the economic opportunities. Many such distributors either market directly to consumers or engage other sub-distributors. Thanks to the incentive structure, they make efforts in promoting the goods, making up for the lack of mass media in remote parts of the country.

The three examples of Smart, Tiger Motors and Hindustan Lever show that the low-income market segment may be a source for creative service and product innovations. It takes a bit of out-of-the-box thinking. It requires you to see potential where others only see poverty.

Overcoming the second hurdle: 'Asian customers are unresponsive'

In Asia, companies find it difficult to get customers/potential customers to provide input during the new product/service development and improvement. Even when the product is developed, companies find it challenging to find early adopters. We believe it is due to a combination of the following four reasons:

1. There is not yet a culture in Asia of customers providing inputs. For various sociopolitical and historical reasons of authoritarian regimes, racial and religious tensions and misuse of information, many Asian customers tend to be secretive and participate less in surveys or providing feedback.
2. Asian companies themselves tend to be arrogant in assuming that they know what is best for their customers. They make little or no effort to try to understand customers and evoke responses.
3. Given that there are many products/services available in the market from both Asian and non-Asian companies, customers in Asia find it unnecessary to spend their time providing input to the Asian companies. While it is true that many of these products and services may not exactly address their needs, Asian customers have become used to adjusting to the available products because very often they believe that they cannot dictate much, given the price they are willing to pay.
4. Being an early adopter means taking risks. Asian customers tend to be relatively risk averse and spend money cautiously. Many items they buy are those that bring social recognition to them. Thus, being 'lonely' with something new is not considered socially valuable by many Asian consumers.

To get around these issues, there are some interesting strategies that some companies have employed successfully.

Engaging the customers for input

For a company to get genuine feedback from customers, it is important for customers to feel that their feedback is respected and wanted. A simple thank you note might in some cases encourage customers. However, there are more creative ways.

iRiver of South Korea is one of the world's most successful independent companies in the design and production of MP3 players. It has created an uncensored bulletin board on its website where a customer can express his or her opinion, criticism, praise or suggestion. Keeping it fair and transparent assures customers that iRiver takes them seriously.

Customers can request certain features, and iRiver, whenever possible, includes those features in the nearest possible upgrades of its products. To keep users engaged, iRiver offers free upgrades of the software for its MP3 player more than four times a year. Each of these upgrades is based on the suggestions and participation of iRiver customers. In this way, iRiver, which now has customers not only in South Korea but around the world, actively taps into the brains of its most sophisticated customers. And guess what? The efforts made by iRiver are perceived to be sincere by its customers.

On the user's forum on the same website, the customers have got so involved that they reply to the questions of other customers. The bulletin board has become a knowledge repository and people share experiences. Now iRiver has opened an iLounge in Seoul to bring customers together and create a sense of community among them and a closer relationship with itself. Heavy users of MP3 players enjoy meeting each other and exchanging experiences and music. iRiver has definitely created an upward spiral of customer involvement.

In its quest to make Singapore a learning nation, NLB has transformed its libraries.[11] NLB brought customers and their needs to the centre of the organization. Staff were retrained and processes were re-engineered and supported by good IT systems. Users of the libraries now have access to the latest books, music, films and documentaries and they can watch/hear them using the library facilities. Unconventionally, there are cafés inside the library where readers can browse through books. The catalogue and reservation service is available online. The books can be sent via post and can be returned at many convenient points throughout Singapore. To avoid queues, customers can borrow books themselves with an advanced kiosk. The libraries are situated in shopping malls near housing estates and metro stations for easy accessibility.

The fact that the complaint to compliments ratio was 1 to 17 in 2002 is an indicator of how well NLB has incorporated the customer input. The fact that customers go out of their way to give compliments is a testimony of their involvement.

Finding early adopters

Finding early adopters is a real challenge in Asia. Often, products and services that are bought have a social value, that is, they reflect on the person. The more established a product/service is, the greater social comfort it brings to the person using it. In such an environment most people are averse to trying something new. Talk about a challenge for innovative companies. Companies should attempt to create environments in which customers will not feel uncomfortable being a first adopter. One way of doing this is to appeal to consumers' deeply held concerns such as those about health or safety.

HLL of India is an interesting example. It has marketed salt successfully for the rural segment of the population on a health platform instead of the traditional 'pure salt' image. With its advertising and consumer education efforts, it targeted young mothers from the lower strata of society who are responsible for their household cooking and purchasing decisions.[12]

Iodine deficiency disorder (IDD) is a disease that causes retardation and lowers IQ. Globally salt is the vehicle used to provide adequate iodine to humans, but in India only a small percentage of the salt is iodized. People are unaware of the dangers of iodine deficiency and often don't know that the Indian style of cooking destroys a large amount of the iodine present in salt. HLL's initial efforts to enter the branded salt market to outshine established national players had failed. HLL had begun to think that differentiation in a commodity like salt was difficult, other than on the basis of purity (which was already the established paradigm of competition). In 1997, however, when the IDD issue was highlighted by the Indian government and the UN, HLL jumped on the opportunity to appeal to consumers on an *iodized* platform. Even though many other companies had iodized salts, their inability to sensitize consumers on health grounds has made HLL's 'Annapurna salt' a clear market leader in a short period of time. HLL came across to the consumers as a trusted company that cared for its customers. Since then, HLL has also marketed other salt brands for the upper strata of society with the right mineral balance – iodized but with lower sodium to help to manage high blood pressure, which is a common health problem in India.

Patkol of Thailand created a tube-ice market in northeastern Thailand in a similar fashion. One day, a newspaper reported that the health department had found harmful germs in block ice. Patkol seized the moment and invited the press to talk about the tube ice which was produced in a sanitary fashion, very rapidly making itself a success story.

Tiger Motors of Thailand is again a quite insightful example. The company had developed a new motorcycle for teenagers that could do no more than 100 km/hr.[13] Motorcycle accident rates are high in Thailand. The founder of Tiger Motors attributed the company's decision to offer this motorcycle to its commitment to social responsibility and to fulfil the needs of the society. This important social reason plus the motorbike's lower price tag pleased many parents, who had to foot the bill. To attract young owners, the motorbike has a stylish look and a trendy brand name – Ozone. There is a pocket in the front console of the motorbike for a mobile phone, the windshields can be changed and come in a variety of different colours, and under the seat there's a plug-in point for recharging the mobile phone battery.

Thus, with its appeal of being socially responsible, Tiger's Ozone attracted public attention, especially that of parents. The lower price tag further enhanced the interests of parents. Excited by the cool features, the young owners would have been open to trying Ozone in any case, but the support of their parents certainly made the case much stronger and more plausible.

Another way to encourage early adopters is to look at the unserved or underserved markets, which tend to be more open to try new things.

A large percentage of Indian farmers are poor, illiterate and heavily dependent on intermediaries who often cheated them by offering lower prices than they deserved. The Indian conglomerate ITC, in its quest to get better procurement deals for soya beans, established e-Chaupal. A chaupal is normally a traditional place in Indian villages where the villagers come together for social purposes or to discuss important matters. e-Chaupal is essentially a computer connected to the Internet via a telephone or satellite connection, and operated by a farmer trained by ITC. A single e-Chaupal serves about 600 farmers living in about 10 villages within a radius of 5 km. With e-Chaupal, the farmers can keep track of prices in the local, regional, national and international markets and decide if they want to sell to ITC, who then do an electronic weighing (unusual in traditional marketplaces where manual weighing is still prevalent) and offer incentives for high quality. The farmer who operates a given e-Chaupal aggregates the demand to get an even better deal from ITC. And the farmers in the community give him a small commission and huge respect for his work. This level of transparency is unprecedented for farmers, helping them to earn more, while getting advice from ITC on how to become more productive. Having established an infrastructure across the villages where more than 70 per cent of the Indian population lives, ITC is also exploring offering credit, insurance and a whole gamut of services and products that the average small farmer/villager has never had access to. ITC has indicated that the US$3000–6000 that they originally invested in each e-Chaupal was earned back within a year and the venture is profitable. In early 2004, the e-Chaupal network served 18,000 villages, reaching up to 1.8 million farmers.[14] And in 2004 e-Chaupal won the inaugural 'World Business Award' instituted in support of the UN's Millennium Development Goals.

Communication is obviously key in finding early adopters. Appropriate channels should be used when targeting early adopters in particular. Within the early adopter groups, the trendsetting/influential consumers are important. Enlisting these trendsetters as your communication allies can be a successful technique. Let us come back to the example of iRiver of South Korea. It went to influential opinion leaders to ask them to evaluate its products on an independent basis and post the results to various websites and chat rooms where the early adopters could access this information. That, the company feels, has been a major success factor behind the rapid adoption of iRiver gadgets. It enlisted its lead users as co-creators.

Netizen Funds is a good example of where a feeling of co-creation has led to increased viewership of a given film. Netizen Funds was an experiment in South Korea to use the internet to solicit film fans to contribute to the production costs of movies in which their adored actors would play a main role. In this way film fans could influence the frequency with which they could view their idols on the screen. There was also the upside of sharing the profits. Such customer involvement created excitement among them, leading to higher film sales. This is something that has not yet been fully explored in other countries, but is an idea that has international appeal and application.

It is obviously important that companies need to create the complementary infrastructure so that customers can easily obtain the products and services thereafter. Patkol of Thailand understood this very well. It knew that demand for tube ice was a dormant demand, that is, one can live without tube ice if the beverages that are offered are cold. In the early days, Thailand consumed very little tube ice. Thus, the distribution was primitive and the suppliers were not active. They wanted the hawker centres to go and buy the tube ice. Since hawker stall owners did not want to put in extra effort for something which was not really in demand, they never bothered. When Patkol got into the business of making ice machines, with its partners it made sure that it could deliver ice on call. The company made well-packed, clean tube ice available to all hawkers who sell food.

What the examples of HLL, Tiger Motors, e-Chaupal, iRiver, Patkol and Netizen Funds show is that one can enlist the traditional conservative Asian consumer to become a partner in innovation. A business must compensate for the fear that being out there as an early adopter is socially less acceptable by playing on other deeply embedded needs such as health and security or by catering to forgotten customer segments. This has to be backed up by a clever communication policy, often involving the early adopters, and by guaranteeing the availability of the product or service.

Overcoming the third hurdle: 'The lack of reliable marketing data in Asia and the heterogeneity of Asian markets'

To bring an innovation to market, companies need some information on the market. Unlike the industrialized world, emerging Asian countries do not have much reliable market data to make business decisions. Similarly, the challenge of heterogeneity in and across Asian countries can be formidable. This has been experienced before by Western multinationals introducing Western products across Asia.[15] Now, Asian corporations are realizing how challenging this issue could be. Under such circumstances, companies either give up and choose not to bother bringing new products to market, or they do not bother to think about ways to get around these two important problems, and go for trial and error. But errors can be big and detrimental to the survival of an organization.

MyWeb was a promising start-up in Singapore that designed and sold set-top boxes for access to the internet. The idea was simple and seemed attractive: rather than buying a PC to gain internet access, consumers could buy a much cheaper electronic box that would allow them to use their TV monitor as the interface with the internet. The implicit assumption here was that many households did not really have any use for the computing power or other applications of a PC, and simply wanted access to information via the internet. Why bother to pay for a PC that you don't need? The company did not have the resources to do a thorough market study before launching the product. It went ahead with it in Singapore and Malaysia and, very soon after, in China. It had some success in Malaysia (basically because of the lower price compared to the PC), but failed completely in its other markets. At the launch of the product, they discovered a lot of new information. It appeared that the TV set was not, after all, the best interface with the internet: for example there were conflicts in the family between those who wanted to surf the net and those who wanted to watch TV. Moreover, many websites require a higher resolution of the screen than that available on a traditional TV set. So consumers did not have access to all sites and felt somewhat cheated. And while consumers objectively did not need PCs, they still wanted them because it was partly a status symbol and partly an investment in their children's education. Many of these insights could have been obtained through a bit of creative market research. And the difference between Malaysia and Singapore in the initial take-up of the product was perhaps predictable. MyWeb had to change its business model and disappeared from the market.

There are no easy answers to the challenges posed by the lack of market data and the difficulty of translating knowledge about one market in Asia to another one. However, based on our discussions with companies, we find three broad groups of solutions that successful companies have found to reduce the risk introduced by the problem of unreliable marketing data and market heterogeneity in Asia.

The first approach is simple: *share the risk* created by the lack of information to reduce that risk. When ChoLam, the Thai clothes designer and manufacturer, decided to go into business, it had no idea how the market would perceive it or even how to run a business. But it had design skills which it was able to sell to the prominent department store of the Central Group of Thailand. Its garments were labelled as that of the Central Group, and it learned the business. This way ChoLam offloaded the risk to the Central Group as long as it was not sure how to manage the market risk. However, as soon as ChoLam felt comfortable that it understood the market better, it decided to go it alone.

ChoLam also quickly learned the important lesson that the greatest value is ascribed to those who own the customers. When the Thai baht collapsed in 1997 and nameless outsourced manufacturing became competitive, the company bravely decided to invest in building its own brand. It wanted to avoid dependence on others. By that time, it had learned the ropes of the business from Central Group.

For every new collection, ChoLam invites its distributors from various countries to its headquarters and holds candid discussions about the possible success of each new design. Distributors find this opportunity useful as their input is often used to modify current designs and develop new ones. The distributors have all been in the business for some time. Having their own self-interest at heart, they try to provide their bit of non-codified knowledge in accepting or rejecting the current designs and influencing future designs.

Sometimes, it is also possible to use the partner's brand to boost one's own image. iRiver has done that effectively by prominently displaying their 'Designed by Inno' label on its products. Inno, which is a design house based in Silicon Valley, is one of the leaders in the field of designing electronic gadgets among other things. Many of the world leaders in consumer electronics use Inno's services.

A second way to reduce the risk is to *go to new places* where one does those things that one knows very well, in order to supply an as yet non-existent market. Many times businesses are not successful because the complementary services that make the product viable are not available. Patkol had that problem when it started exporting ice machines to Malaysia. There had been no demand and consequently no delivery support services were available. To make sure that its ice machines would be attractive to the ice suppliers, Patkol created services like delivery mechanisms and so on. As the CEO of Patkol said, 'Today, we are not exporting ice machines. We are exporting the ice business. It is different. That's our marketing technique.'

The third strategy is to develop a system to *learn quickly through trial and error*. Being quick is essential here. Carry out many small experiments and learn from them. Lapid Foods of the Philippines is a small chain of high-end, small, local snacks called chicharons. Its CEO used some ingenuity to solve the problem of not having any market data on snack consumption. The problem was how much raw materials for chicharons should be kept ready so that it could offer fresh, top-grade products, which was the key promised differentiation in the market. There was no way to ascertain the number in the beginning – government statistics about people living in the area were not only difficult to obtain but also unreliable. The CEO, a lawyer and accountant turned entrepreneur, used some creativity. Before he set up the first outlet, he counted the vehicles passing, noted the type, the ratio between the public and private vehicles and the number of pedestrians. The technique, although imperfect, has served him well, and he used it every time he opened a new outlet. His customers are very happy with the freshness of the products and his inventory is under control.

Neowiz, which in 2004 was one of the top 20 companies by market capitalization in South Korea, is the leading player in online games and the avatar industry. Its success depended on how well it had understood the nuances of Korean culture and integrated them into its offerings. It was, however, difficult for Neowiz to replicate its success in the same industry in other countries. To try to adapt avatars for Japan, Neowiz has set up a new entity in Tokyo where young Japanese developers work under the direction of South Korean managers. But Neowiz knows that it's going to be tough. It will prototype for a year in Japan. For China, it has studied who can be the local partners, necessary because of cultural and regulatory factors. If nothing happens in Japan, it will think carefully about its future plans.

The issues of unreliable marketing data and market heterogeneity have no simple answers. The degree of adaptation of form and content depends on the industry and the country. While significant adaptation might be required for clothing, consumer electronics require less adaptation. Being aware of these differences is a good start. And we hope that these examples have shown some successful solutions to overcome the lack of data and market information.

CONCLUSION

We started this chapter by saying that innovation without intimate customer knowledge is going to be difficult. We know that Asian markets are not the most supportive for an innovator. Customers tend to be more conservative, markets are heterogeneous, market data is not available and the most sophisticated markets are far away, be it geographically or culturally. But the examples of companies like Patkol, Hindustan Lever, Smart Telecommunications, Tiger Motors and others suggest that there are creative ways to overcome the hurdles and challenges. Some suggestions we hope you will retain from these inspiring examples are summarized below.

Sophisticated markets can be an important source of learning. Some innovators will set up antennae to tap into these sources. There are several approaches possible, ranging from the creation of a wholly owned subsidiary, setting up an alliance or acquiring an overseas unit. None of these strategies is easy to implement, but the examples of Haier, AmorePacific and Samtel can be inspiring.

Asian markets are often considered to be poor and thus uninteresting for innovators who want to launch new products and services at premium prices. But these low-income markets also tend to be large. Smart Communications and Tiger Motors are two examples of companies that have used their creativity to customize innovation for mass markets instead of just focusing on the crowded premium segments. Size can compensate for premium prices.

It may be tough to find early adopters for cultural or other reasons. But as the examples of Hindustan Lever, iRiver or Patkol show, it may well be possible to use indirect ways to seduce lead customers to adopt the innovation and become your best ambassador. Backing this up by a clever communication policy, often involving the early adopters, and guaranteeing availability of the product or service can help.

Finding marketing data is tough, so experiment. Heterogeneity needs to be overcome. Awareness of this lack of data and the differences between Asian countries is the first step. After that you can find ways to engage customers, share the risks created by the lack of information and set up careful experiments to learn.

Notes

1 This example draws significant information from Min K.J., 2004, 'Sweet Smell of Success' *Far Eastern Economic Review,* 18 March.
2 Shiseido of Japan has successfully adopted a similar strategy in the past. This is documented in Doz Y.L. and Kazuhiro A., 1998, Shiseido France, INSEAD case study.
3 Wu Y., China's Refrigerator Magnate, 2003, *The McKinsey Quarterly,* (3).
4 Hong Kong Retailer Giordano Pulls Plug on Europe Venture, *The Asian Wall Street Journal,* 3 October, 2002.
5 Je ne Texte Rien, *The Economist,* 7 July, 2004, p. 57.
6 Williamson P., 2000, China's Beer Industry, INSEAD case study.
7 For a more elaborate discussion on this matter, please refer also to the cases in Prahalad C.K., 2005, *The Fortune at the Bottom of the Pyramid,* Wharton School Publishing, Philadelphia.
8 Phoning Home Pesos from Across the Ocean, *International Herald Tribune,* 9 August 2004.
9 Crispin S.W., 2003, 'Tiger Bikes Do A Roaring Trade', *Far Eastern Economic Review,* 9 October.
10 Tiger Roars on to Motorbike Scene, *The Nation,* Thailand, 19 December, 2003.
11 De Meyer A. and Garg S., 2004, National Library Board of Singapore, INSEAD case study.
12 Annapurna Salt: Public Health and Private Enterprise, Case study by Michigan Business School. Reproduced in Prahalad C.K. op. cit.
13 Tiger Offering Slower Rides, *The Nation,* Thailand, 15 December, 2003.
14 Waldman A., Internet Transforms Farming in India, *New York Times,* 2 January, 2004. See also Prahalad C.K. op. cit.
15 Dawar N. and Chattopadhyay A, 2002, Rethinking Marketing Programs for Emerging Markets, *Long Range Planning* **35** 457–74.

6 Mobilizing resources

In Chapter 4, we discussed how organizations in Asia need to and can rejuvenate themselves for the new competitive realities. In Chapter 5, we talked about the challenges pertaining to markets for innovation by Asian companies. But, to *create* new products and services and *market* them effectively, firms need certain *resources*. We discuss the issues related to the resources for innovation in Asia in this chapter. The discussion in this chapter has relevance to all innovators but may be particularly important to those who find themselves similar to the respondents in cluster 3 of the typology presented in Chapter 3, the ones who feel that they are poor in knowledge resources.

One of the key challenges for innovation in Asia is that it is a resource-starved environment. We are not referring to natural resources, which are abundant in many parts of Asia. In fact, some scholars now think that the abundance of natural resources in many Asian countries diverts the attention of policy makers and businesspeople from the need to move towards more knowledge and innovation-driven development in those countries. The discussion in this chapter will focus specifically on talent and capital. We introduce this by briefly touching on the need for an adequate technical infrastructure.

TECHNICAL INFRASTRUCTURE

Innovators need to rely on a good technical infrastructure. Companies in Malaysia, Singapore and Korea, who specialize in information and communication technologies (ICT), benefit from a performing broadband network. Heavy industry requires a steady and reliable supply of electricity. The rationing of electrical power in cities like Shanghai in the summer of 2004 shows how fragile economic growth can be when the capacity of the basic technical infrastructure is insufficient. Innovative traders like Li & Fung require working ports. A reliable, operating technical infrastructure is a necessary condition for innovators to blossom.

In a way, one can also see the educational system as a part of that infrastructure. The network of high schools, technical institutes and colleges and universities can be seen as an infrastructure that provides one of the key resources needed for innovation: creative brainpower.

Table 6.1 Technology infrastructure indicators in selected countries

Country	Gross enrolment in tertiary education % of population (1999)	PCs per 1000 persons	Telephones per 1000 persons	Internet users %	Scientists and engineers in R&D per million people	Electricity use in kwh per person
China	6	28	328	59.1	545	896
India	11	7	52	16.5	157	365
Japan	37	382	1194	55.9	5095	7237
South Korea	43	556	1168	26.7	2318	5288
Singapore	44	622	1258	2.2	4140	7178
Hong Kong	28	422	1507	2.9	93	5541
Malaysia	13	147	567	7.2	160	2731
Thailand	35	40	365	4	74	1508
Indonesia	11	12	92	5.3	na	404
Vietnam	9	10	72	0.6	274	325
United States	45	659	1134	65.8	4099	11714
United Kingdom	45	406	1431	34.3	2667	5652

Source: The World Bank (2003) UNESCO (2003)

Table 6.1 brings together some key indicators relating to the technological infrastructure of some selected countries. It includes the USA, the UK and Japan in order to give some benchmarking information. Gross enrolment in tertiary education gives an idea of the potential available brainpower. The number of engineers per million of the population is a similar indicator. The number of PCs or telephones (fixed and mobile lines) per thousand people as well as the percentage of internet users give an indication of the potential for ICT innovators. Electricity consumption per person (which in fast developing countries is often close to the available capacity) can be an indicator of the extent to which a country can go into heavy industry or services requiring air-conditioned offices.

Needless to say, the differences between the countries are great. The number of engineers per million people or the electricity consumption differs with an order of magnitude between China and India on the one hand, Japan and the USA on the other.

It could be said that, in absolute terms, the total number of engineers or the power-generating capacity in China and India is still large. But, those countries need these resources in proportion to their geographical size and population to run the 'operations', for example transport, telephone systems, construction and so on. And only a small number of these engineering experts are available for innovative and creative work. If the proportion of technical resources compared to the size of the country is low, then that small number available for creative work will become very small indeed. Unless some of the countries cited in the table invest heavily in educational and technical infrastructure, they will find that technical talent will become scarce and thus expensive. This is already happening in the coastal cities of China. We will come back to the importance of engineering skills in particular in the next section.

Managers may argue that there is little they can do about this. Putting in place infrastructure, such as a fully operating electrical production and distribution system, a vastly expanded education system, or abundant telecommunications capacity, remains to a large extent the initiative of the government. But innovative entrepreneurs can do three things: vote with their feet, lobby for investment in infrastructure and provide innovative solutions to the governments.

Voting with your feet, that is, taking your business where there is available infrastructure, may look opportunistic. But that is the free market. If innovative entrepreneurs take their business where they find the technical infrastructure, governments will take notice and perhaps react to it. Lobbying can help and it may well be the corporate social responsibility for innovative companies to call upon governments to make the appropriate investments. But the most important message is that investment in infrastructure offers a huge opportunity for innovation. One example of this is the creation of schools and institutes of higher education to respond to the need for skills and talents. Some Indian companies, for example NIIT, are now among the largest and most international providers in the world of high-quality IT education. Education is a booming business that can be financially rewarding.

HUMAN RESOURCES

Human resources are the most fundamental of knowledge resources that are necessary for innovation. We believe that the currently available human resources are inadequate to provide effective support for innovation efforts in Asia.

It is true that these resources are vital for innovation everywhere. And, even in the US (which arguably has, on average, the highest number of innovative companies in the world), there is always a hue and cry from firms about the lack of skilled human resources. However, two factors make the scarcity of skilled human resources more pressing in Asia. First, in industrializing Asia, these resources are below the *hygiene* level, and, when present, they are inaccessible to those firms that want to innovate. In Chapter 3, we made the point that there is a large group of companies unable to innovate due to a lack of knowledge resources. Specifically, good skills in engineering, design and general management are scarce in Asia. Second, it has been difficult for industrializing Asian countries to develop skilled human resources or attract and retain skilled foreign human resources. Human capital formation takes a significant amount of time. Attracting foreign human capital takes significant financial resources which most Asian firms lack.

The next sections examine the impact of possible shortages in engineering, design and management skills and the general supply of creative employees.

Engineering skills

Think, for a moment, about BMW, Mercedes or Toyota, Hewlett-Packard or Dell, Intel or AMD, and GE or United Technologies. Can you imagine any of these companies, known for their commitment to innovation, being where they are now *without* top-class engineering skills? The companies themselves cannot imagine this either, and that is why, even today, they have a relentless focus on engineering.

Engineers make things happen. They are the craftsmen, so to speak, who realize the dreams of strategists and R&D directors by making appropriate trade-offs in product development. Thus, without good engineers, many innovations cannot be realized.

Let us come back to the issue of availability of engineers, introduced in the previous section. There is a general impression that Asia has plenty of engineers, but the numbers indicate otherwise (See Table 6.1). What we see in that table goes against common wisdom. Asia *lacks* engineers in R&D! Indeed, we could ask whether it is the absolute number or the number of engineers per capita that matters more. While we do not deny the value of absolute numbers, it is important to look at the average quality as well. In our research, we often encountered the complaint that engineers in Asia – often even those coming from the top universities in the nation – are not *practical* enough. The fact that young Asian students consistently rank amongst the top in the world in science and mathematics, but later do not become creative engineers, suggests that they are unable to translate their theoretical knowledge into practical engineering skills. Thankfully, there are always exceptions. The fact that we hear so much praise about the exceptions, such as engineers from the top engineering schools like the Indian Institutes of Technology in India, Tsinghua University in China and KAIST in Korea, in fact provides further evidence that engineering talent is scarce.

Other factors aggravate the scarcity of engineering skills. Top-quality engineers from many of the Asian countries migrate to economically developed countries in Western Europe, Japan or North America. A sizable number of engineers in Asia also tend to change their careers to more lucrative and socially prestigious non-engineering careers, including accountancy, law, management and consulting. Many good engineers who stay in Asia tend to be lured away by the Western multinationals that pay them better salary packages but often underutilize them: they are required to carry out mainly basic production or maintenance work. As a consequence, Asian firms are confronted with a general shortage of good engineers, which affects their competitiveness and ability to innovate successfully.

Some companies have found their own creative solutions by trial and error. Shin Satellite has one of the most sophisticated wireless broadband technologies in the world, yet it is based in Thailand. How does it gain access to technology skills? Other than focusing on systems engineering and project management, Shin Satellite does not direct its efforts specifically on technology. It sees its role as understanding customer requirements. Indeed, it has also brought foreign talent into Thailand and continues to hire the best available technical talent in the country and further develop it. But, Shin Satellite's key strength is its partners. It has good partners in specific technology areas around the world. They range from some companies in Scandinavia to academics in relatively unknown universities. Its competence is in finding them and managing them. In this way, Shin Satellite manages the scarcity of technology skills.

One could argue that such an arrangement requires deep pockets. So let us return to the example of Aapico Hitech, a smaller company with fewer resources which is also based in Thailand. Aapico invites potential customers and partners to visit its premises and offers to work for them, even though in many cases there is little money involved immediately. The objective is to hone the technical skills of its technicians and engineers who learn 'to listen, talk and do' through these opportunities. It has also tried to get the best engineering talent by offering academic sponsorships to study in the top technical university in Thailand. The CEO has created a sustained focus on engineering in the company by highlighting that, as an automotive supplier, having top-grade engineering is a precondition for survival. Thus Aapico provides incentives to its engineers to always keep learning. Importantly, the CEO encourages them to travel and develop personal relationships with their peers in other companies and countries and learn from them. Aapico's partnerships with various top automotive manufacturers have forced it to create a learning system whereby the new knowledge is absorbed and diffused in the organization rapidly.

Another possible option is to tap into the talent that wants to return to the country for personal reasons after having gained the relevant experience in sophisticated markets. For instance, the Taiwanese semiconductor industry benefited a lot from the Taiwanese diaspora that wanted to return from the US to Taiwan. Many Taiwanese engineers went to study in the US and stayed there and worked for US companies. But after many years, some were homesick and returned. These engineers brought home their education but, perhaps more importantly, many years of deep understanding of product and process technologies. In different ways, Indian IT services companies are also using the talent of overseas Indians, but more for business development purposes.

Design skills

Equally important for innovativeness in Asian companies is the ability to *design*. Designing is a highly skilled job: it means the ability to envisage a complete product or service and then create an architecture that engineers could use. Aside from the electronics sector where some Asian companies have a lead and have begun to design middle and even high-end products, design capabilities in Asia in other important sectors such as apparel, automotive or chemical engineering are largely underdeveloped. Even in software development, Asian companies are not known for their design skills but mainly for their production, that is, relatively low-end programming skills.

To a large extent, the lack of design capabilities is because Asian companies have been essentially original equipment manufacturers (OEMs), supplying the big principals. In most of these relationships the design was not their responsibility. However, the competitive field is changing. As discussed in Chapter 1, China offers a price/quality combination that is difficult for anybody else to offer in manufactured goods. Therefore, the formerly low-cost suppliers in Southeast Asia have lost much of their attraction. Consequently, they are forced to move up the value chain. They can do this by either designing for the branded goods specialists in the industrialized world or introducing their own innovative goods and services in the market directly. Either way, design skills are critical for them. Today, in a large number of industries, it is the design skills that allow firms to create and capture significant value and margins. It is not surprising, for instance, that Nike has chosen not to do any manufacturing of its own for a long time and focuses on design.

Your challenge is how to *develop* these design capabilities. As we said in Chapter 1, this has been the most difficult thing to learn in the process of going up the value chain. Asian firms rapidly graduated from simple assembly in the 1960s and 70s to process improvements and developments in the early 1990s, and it seemed likely that they could progress to threaten even Japan.[1] However, since then, their achievements and attempts in the next stage of being able to design have been rather limited, other than in the field of consumer electronics. There are few design institutes of any kind in Asia and where they are present, the quality is not up to international standards. For instance, until a few years ago, Thailand, which is well known internationally for attractive textile products, had no textile design schools. Even today, those who can afford it still prefer to go to Europe to learn about design because there is a higher chance of success in the future. The situation has begun to change slowly. Some emphasis is being placed on design in existing study programmes. And, there are new programmes focused solely on designing in the capitals of Southeast Asian countries. However, unless the trainers are trained themselves, equipped with the right tools and techniques and given a chance by customers, progress may not be fast enough.

The Indian IT industry offers interesting insights into the design issue. This industry is strongly focused on the export of IT services rather than products – services represent more than 75 times what is earned by Indian IT companies for their products in 2003.[2] This bias towards services is created by the uniformity of available skills, aggravated by a shortage in skills such as programming design and specification in engineering, which are required to a lesser degree in service businesses. Such design skills are absent due to distance from the sophisticated markets, namely the US, for IT products. This makes funding and further learning opportunities for design somewhat limited.[3]

Designing requires creativity, planning and focus, a complete view of various associated elements, an understanding of customers and most of all self-confidence. By and large, the education system in Asia encourages memorizing instead of thinking creatively. As far as planning and focus are concerned, Asian managers and employees often tend to prefer *improvisation*. Improvisation is a good thing and should be encouraged. But relying on it is not something that can bring about sustainable and reliable results.

Being OEM players for certain components, most Asian companies lack the complete picture of products and the complexities involved. That is difficult to master! Illustrating this point, Takahiro Fujimoto, a prominent professor from Tokyo University, recently argued that Japanese manufacturers' real strength is in those 'products that require many components to be designed in careful detail and mutually adjusted for optimal performance'.[4]

We have discussed the issue of understanding customers in Chapter 5. Research shows that few, if any, original design manufacturers have the consumer knowledge and marketing expertise they need for success in developed markets.[5] And, designing for local Asian markets with low spending power requires a different mindset.

Self-confidence arises from accomplishments. Given that Asian companies have had limited past accomplishments in designing, they have low self-confidence. This is discussed in more detail in Chapter 8.

Even if the Asian companies break the challenge of design capabilities, they will find it difficult for their designs to be *accepted* in domestic and international markets. In fact, without being exposed to the latter, designers often find it difficult to create something that might be truly appreciated and adopted in home markets. This seems to be the experience of a large number of Asian apparel designers in Asia.[6] When there *is* limited and early success, the designers are short on business skills. In many Asian societies, even today, Western goods are preferred because of the high social status that they apparently bring to Asians in Asia. We discuss in detail these crucial perception challenges and their negative impact on confidence in Chapter 8.

What do you do about this? Realizing the deficiency of design skills in South Korea, iRiver knew it had to do something different. Soon after iRiver entered the MP3 player market, it was clear that the market would be flooded with competing products. The CEO knew that even though iRiver's products were technically superior, iRiver needed to establish an emotional link with customers. Only a lifestyle approach would ensure long-term success in the market. It was clear to him that iRiver needed a good designer so that the product would be aesthetically appealing. He contacted Inno, a Silicon Valley-based leading designer for consumer electronics goods, that advises a large number of companies. iRiver told Inno that the company had little cash but a bright future which Inno could assess, being in contact with all the leading consumer electronics goods companies. They signed an unusual agreement that saved iRiver the scarce cash and offered Inno a piece of iRiver's future. With its ambition and willingness to ask for unusual agreements, iRiver could tap into one of the best designers in the world. Even today, Inno advises iRiver proactively on new designs. Each iRiver MP3 player carries the distinctive label 'Designed by Inno'. As iRiver now goes to more countries, it realizes that it will need to design its products a little differently for different parts of the world.

Aapico, on the other hand, has invested in learning and honing its design skills. In 1991, Aapico became the first among its peers to introduce CAD/CAM technology and CNC machines. Aapico learned designing. Until the mid-1990s, Aapico made jigs from German, Australian and American drawings. It felt constrained and decided to have its own designs. The CEO recruited an experienced and well-travelled Australian designer and engineer from Ford in order to set up an in-house jig-design and engineering facility. The designer loved Thailand and loved to teach. He has trained a group of people to be designers within Aapico who have been retained by Aapico through various incentives and fair treatment. Every Aapico employee is a shareholder, a policy initiated by the company when offering shares to employees was still rare. Today, the company claims that it has the best concept design for low-volume tooling in the world. On price, quality and design it can compete with anyone.

Aapico's success has not gone unnoticed in the auto-parts business. The company diversified into consulting when other firms expressed an interest in learning from it. Along with DaimlerChrysler's Lean Manufacturing Center (LMC), Aapico advised the Indian auto-giant Telco (part of the Tata group) on jig design, for instance.[7] In the process, it continued to learn from DaimlerChrysler.

Design is not necessarily always associated with technology-intensive industries. Products and services can be designed using the *indigenous* knowledge and skills. The example of the Martha Tilaar Group from Indonesia offers some insights. Its ambition is to be a leading cosmetic company with an Eastern and natural emphasis but utilizing modern technology and R&D. And it has had reasonable success. The company's founder Martha Tilaar was an exponent of the value of traditional Indonesian knowledge and the *jamu* herbal mix. The company embraced the legacy of the ancient Indonesian beauty philosophy known as *Rupasampat Wahyabyantara*, which means 'Beauty is an expression of a harmony between the outer and the inner'. The outcome of this philosophy is the Sari Ayu cosmetic range. Following the initial success, the company created a range of products, such as Biokos Total Age Care, Belia, Dewi Sri Spa which combine natural Indonesian ingredients with modern research and production techniques and strict quality control, for which it has gained ISO quality approvals. Over 33 years, the initial venture has grown into a group of companies known as the Martha Tilaar Group, employing over 6000 people and producing cosmetics, herbal medicines and food supplements, as well as operating several beauty schools, salons and spas in Indonesia. The Group's activities are based on the foundation of indigenous knowledge, and its market leadership is a result of its own design efforts.

Similarly Eu Yan Sang of Singapore is a leader in oriental medicine in Southeast Asia. Oriental medicine is grounded in indigenous and traditional knowledge over thousands of years, and Eu Yan Sang has developed capabilities to harness that knowledge. Not only that, the company taps into the latest scientific knowledge and focuses on quality. For instance, its manufacturing plants have ISO certifications, it partners with Western natural products companies through equity stakes, works with a Hong Kong university for scientific research on Chinese medicines and co-develops a Chinese herbal product range for spas. Today, the company is in the driving seat as a designer by setting up consulting clinics in Australia and promoting fusion treatments combining Western and Oriental medicinal treatments.

So far, we have focused on the design of products, be it high or low tech. Design also exists in services. Two short examples will illustrate this. One area where Asia has already had a major impact on consumers in industrialized countries is that of food. You cannot go anywhere in the world without finding a full range of Japanese, Korean, Chinese and Indian restaurants and caterers. This is obviously nothing more than simple export. But something interesting has happened in the higher segment of the market. Some of the most sophisticated restaurants in the world, whether in Tokyo, Paris, Singapore or Hong Kong, have specialized in fusion food, that is, the combination of Western and Eastern ingredients in order to create unusual dishes. You could argue that this is still product design. But most of these fusion food restaurants are also developing a different style of interior decoration, serving and so on. Fusion food is more than food, to some extent it has become a lifestyle. The most sophisticated restaurants of this style are still in Tokyo's Nishi Azabu district, but in many Asian capitals interesting experiments with this blending of East and West are to be found.

A second example of design in services is the case of Banyan Tree Hotels and Resorts. This chain of holiday resorts has developed a specific style of resort, called a 'sanctuary for the senses'. The company's commitment to relaxation, respect for the natural environment and so on gives them a specific style and design.

Design skills are critical to successful innovation. Rarely are they explicitly discussed in textbooks about innovation management because they appear to be less important compared to pure research capabilities or a strong marketing orientation. Design is about the integration of the different components in a complex design, understanding customer needs and translating the original innovative idea into a concept that can be understood by customers. It is also about creating confidence that the company can respond to market expectations with great products. The examples we gave from both technology-intensive industries and service companies show that it is feasible for young Asian companies to master the art of design.

Management skills

Engineering and design skills are important. But, without management, innovation would often remain clever invention. Asian companies sometimes lack the high-quality general management skills needed for managing the innovation process. Often, functional specialists are promoted to general managers but lack the appropriate training and perspective. This has been true even in countries like India that are supposed to have more developed management practices.[8] Furthermore, management personnel often lack the experience or skills necessary to internationalize effectively. To some extent, this is due to an inadequate exposure to the sociocultural variety that exists within Asia and beyond. While South Korean companies have seen some success, even companies from some of the more developed Asian countries, such as Singapore, have not been very good at managing their overseas operations and relationships.

From our interviews, we discovered that Asian companies to a large extent lack the sophisticated marketing skills that are essential for innovative products and services. Such skills were not needed in the past era of shortage economies. Even today they are not adequately emphasized and developed. Many Asian companies think of sales departments when marketing is mentioned. Again, this stems from the fact that a large number of companies were mainly involved in trading and their growth in the 1990s, driven by manufacturing, was so good that they didn't need to do any marketing. There is also the lack of another important skill, that of public communications; its implication is discussed in Chapter 8, where we discuss the negative perception conveyed by the international media about Asian firms.

How do you overcome this? Having looked through many of the case studies, we found some interesting observations. A large number of effective managers of the companies we studied have had some kind of extended contact with or exposure to North America, Europe or Australia. Some of them had gone to study in the US, Australia, the UK or Japan. Others had extended work experience abroad when, early on in their careers, they worked for multinational companies. But the experience abroad is not sufficient in itself. It may have provided them with training or broadened their minds. But there is a need to do some 'maintenance', that is, have a regular challenge to one's current knowledge. They somehow needed to continue to benchmark themselves, like an athlete who needs an international competition in order to keep performing at the highest level. This benchmark was often provided through a conscious effort to keep alive those relationships created in the early days. The message is clear: *develop* and *maintain* good personal networks with your peers in the industrialized world. They will keep you on your toes when it comes to management practice.

Education for creativity in the workforce

Engineers, designers and managers need collaborators. If they want to innovate they need creative collaborators. The education systems in Asia need significant improvement in two main areas: availability and quality of creative graduates. With the efforts of national governments and international institutions, the progress is taking place at a different pace in different countries. But, the biggest change that is needed is in the mindset that is shaped by the curriculum. Most Asian universities inadvertently emphasize memorizing instead of thinking. This was the almost unanimous impression of the senior executives we have spoken to over the years. And they probably have a good insight because a large number of them are products of the system. While efficiency was enough in the past, critical thinking is now needed for the success of Asian firms.

Massive changes in educational systems are difficult and take time. Many countries have started changing the school and university curriculums, but it is a difficult and slow process. Singapore is an interesting case study of this process. The changes made in the educational curriculum do affect mindsets and capabilities but it takes 5–20 years to see the results in the society.

This is not encouraging if you want to make your company an innovative company in Asia today. But it is always better to be aware of the reality in order to find a solution. The education system in Asian countries will struggle for some time indeed, but individual companies can find ways for themselves to grow and innovate despite these challenges. And perhaps you can take some actions to stimulate that latent creativity. Once more an example can provide some inspiration.

The National Library Board (NLB) wanted to be innovative. To help Singaporeans create and thrive in the knowledge-based economy, it saw its role as helping to increase the learning capacity of the nation. To start attracting the public to the library, it wanted to reorient the NLB focus from books and librarians to customers (formerly called users at the NLB). It wanted to be customer-centric. The problem – nobody in Singapore or elsewhere offered training programmes to make librarians and library staff more customer-oriented. Traditionally, the librarians never learned much about how to serve the customers, and were taught to focus on the management of the library material instead. The NLB took the step of defining its own requirements and training its own people. It had help from various consultants and designed its own programmes. The programmes have been effective and are looked upon with great esteem in the library world. The customers seem to be happy too. This happened despite the fact that the library staff has had a fairly traditional general education.

FINANCIAL INFRASTRUCTURE

An efficient financial infrastructure is the lubricant of a performing economy. There are many components to this infrastructure, but it is largely defined by governments. Managers have little influence over the financial infrastructure of countries, and thus we are not going to discuss it here. Generally, it is sufficient to say that capital market reforms, including those in the areas of disclosure and governance, are needed and are taking place at a different pace in different countries in Asia. What we will discuss is the issue of the availability of capital, which is a managerial responsibility (even though it is dependent on the financial infrastructure).

Financial resources are important for innovation. There is convincing research that the government-led initiatives that were appropriate in Asia during the early industrialization stage need to be reformed, and that a larger type and number of financial intermediaries should be allowed to flourish so that firms can obtain financing, especially for their innovation efforts.[9]

Corporate financing in Asia is dominated by the banks. Most of these banks are state-owned, state-directed or heavily regulated (for example in China, Taiwan and most of Southeast Asia) and have little international clout or experience. Hong Kong is dominated by a handful of big banks with lots of small banks being affiliated to family business groups and South Korea's banking sector has one or two bright stars but has largely been in trouble since the Asian financial crisis. The Indian banking sector is simply trying to be more efficient with a handful of large private sector banks setting high benchmarks. But it remains by and large dominated by government banks.

Given the nature of debt contracts that limit the upside, banks have little incentive to take risks in financing innovation initiatives. In any case, standard bank loans require collateral of physical assets. This is not easy for companies because the assets that could be used as collateral are often already being used as collateral and the young but high-potential, knowledge-based companies have limited physical assets. In his landmark work *Mystery of Capital*, the economist Hernando De Soto found that a large number of businesses across the developing world do not have all their physical assets legally registered due to various bureaucratic hurdles. The implication is that they cannot access the financial markets at all. In any case, a company's collateral against a bank loan is often meaningless because the market for auctioning off the collateral does not exist in Asia. But, even for large corporations, getting loans has been difficult in many countries in Asia, because the Asian banking sector is believed to have been in trouble due to them. After the Asian financial crisis, the banks have been trying to refocus on consumers instead.

Many governments in Asia have set up special banks to support the innovative activities of SMEs but they are either more or less dysfunctional or they are too slow. Very often the banks are expected to focus on certain priority sectors, to the dismay of companies in other sectors. Bond markets do not exist in many countries and, even if they exist, they are not liquid. This is true even for a large country like Indonesia.

As for the equity markets, other than India, most of Asia does not have an equity culture. Given the complex cross-holding structures, limited minority shareholder protection and a lack of transparency, investors find the equity market too risky and participate for trading purposes without understanding the companies and their businesses. Even to be listed, a firm needs to be a significant size, although it is different in NASDAQ-style exchanges for young companies in Singapore, South Korea and Hong Kong. Issuing American depository receipts is possible but requires high standards that a large number of firms are unable to meet. Private equity and venture capital in Asia are limited.

The governments (such as in Singapore and Taiwan) have tried to play a role in venture capital but with limited success. Both private equity and VC markets are underdeveloped because:

1. It is difficult to develop new key products or services in Asia for sophisticated markets far away
2. Innovations for local markets either do not require large-scale investments or the investors have little ability to appraise their importance to the local markets
3. Exits through sales to larger companies or listing on stock markets are less likely for Asian investee companies.

And, even when Asian companies obtain investment for their innovation efforts, their cost of capital is quite high, reducing their competitiveness in international markets. Foreign capital is highly unpredictable (as was seen during the Asian crisis) and FDI is sometimes restricted in many Asian countries due to regulatory frameworks.

How do you innovate in an environment where capital, especially risk capital, is difficult to come by? Even though a firm may have all the engineering, design and managerial resources, without risk capital no innovation can be realized. The good news is that we saw companies pursuing two simple but effective paths to find risk capital. The first approach is an obvious one: reduce the requirement for capital. The second is almost as obvious: look for capital elsewhere and do not rely on traditional resources. Simple as these two solutions appear, most companies tend to underestimate their value. We have seen a few companies in Asia that have overcome their capital requirement issues by coming up with some creative ideas that essentially fall into one of these two categories.

Let us first look at alternative sources of financing. Patkol, the producer of ice-making machines has reinvested its internal cash flows to avoid looking outside to banks for funds – as many other companies do. Of course, this meant that it had to focus on making its customers happy. In fact, its customers were so happy that Patkol's ice-making machines were called 'money-making machines'. The customers were willing to pay one year in advance to order a machine. These pre-payments have significantly reduced the need for equity.

In a similar way, Aapico reinvested most of its cash flow back into the company to sustain and grow its operations. At the same time, Aapico continued to raise bank loans using its good reputation with banks that it had developed both in and outside Thailand. Aapico regularly serviced the loans despite the tough time it faced during some years of its operations, further developing its credibility with the banks. Consequently, Aapico could raise some capital at the time of the Asian financial crisis when its competitors and the lending banks were in trouble. As it had good social and professional relationships with many automotive suppliers in many different parts of the world, Aapico could minimize the negative impact of the Asian financial crisis on its business.

We observed that companies also ran various simple operations to fund their innovations where venture capital was difficult to come by. For instance, iRiver had an engineering workshop. This was a sustainable business by itself and was made easier to keep going because its founder had years of experience as an engineer with Samsung. iRiver reinvested the earnings from the engineering workshop to fund the development of the MP3 coder-decoder that led to its famous MP3 players. Furthermore, the engineering skills in the workshop could be used for the MP3 codec business later.

Government funds may not be readily available, but it is sometimes possible to sell the innovation in a way that resonates with their sociopolitical objectives in order to gain government support. VazBuilt from the Philippines, which we will describe in more detail in Chapter 7, did exactly that. It could provide low-cost, prefabricated housing. Saying that its innovation would help the poor to live with dignity and make the environment more pleasant, the company gained government support and some grants and tax concessions. Grants of this nature are available in many countries in Asia.

Some companies find it better to go to other countries where it is easier to find the capital. iRiver went to a Hong Kong company called AV Concepts.

iRiver's CEO Joon Yang had credibility with them through his former job at Samsung Electronics. iRiver chose to go to Hong Kong because funding for something like an MP3 player company was not possible in the late 1990s in South Korea: the venture capitalists (independent and corporate) were interested only in dot-com businesses. Joon Yang thought Hong Kong was a more open and effective financial centre.

Although an interesting idea, foreign capital always has its own challenges. MyWeb, a set-top box company based in Singapore and Malaysia, had trouble raising money in Asia. So the company raised capital from a Silicon Valley venture capitalist. However, when the company was listed through a back-door listing on the NASDAQ and ventured into China, its US investors did not understand anything about the Chinese market. The investors judged the market with a US mindset and forced the management to take decisions that were detrimental to its success. Finally, the company closed its doors. So, raising capital away from home is not easy, especially when the product or service requires a high degree of localization. If management feels that it will not be trusted by foreign investors, it is advisable to be cautious.

Netizen Funds in South Korea came up with a creative form of involving customers in funding. Until a local movie called *Sheri* struck a cord with the Koreans in the late 1990s, they preferred Hollywood movies. By 1999, the interest in local films was on the increase, with record screenings. Of course, this meant that the production and marketing costs shot up too. Intz.com, a marketing firm, was engaged to market a wrestling comedy called *The Foul King*. Intz.com offered an opportunity to online users to invest in films directly. The intangible that such a model offered to film-crazy fans, especially from the younger generation, was the thrill of helping to make a film of their choice. So the promoters marketed the idea as a means to 'get involved' rather than a tool to gain large returns on their investment. The money raised by Intz.com was small (US$77,500 or so) and was used to fund offline promotional activities and media events, to which investors were invited. A vast majority of investors were young and invested less than US$100 per person. Such investors became unofficial ambassadors of the film, and provided some free word-of-mouth publicity. *The Foul King* opened in early 2000 and was a runaway hit. The film earned more than any other that year (except two Hollywood films) and in the process earned a handsome 97 per cent return for its investors.

The goal of innovating companies should also be to reduce the cost of coming up with the new product or service, thus reducing the need for financial resources. One possible way is to involve customers, as the case of Netizen Funds illustrated. There are some other interesting examples. iRiver partnered with AV Concepts not only because the latter could inject some capital, but also because it had access to good-quality manufacturing facilities in neighbouring Shenzen. iRiver had originally considered investing in manufacturing capacity in South Korea. But considering that all portable manufacturing had slowly moved out of South Korea, iRiver would have needed to invest a significant amount of capital in developing that capability again. Thus, by having AV Concepts as its strategic partner, iRiver reduced its cost of innovation. At the same time, AV Concepts also got higher value out of the deal.

Considering that travel and telecommunication is easier and cheaper than ever before, firms should take advantage of outsourcing work they cannot do efficiently and effectively themselves. Not only does it help to bring the product or service faster to market (remember we said in Chapter 3 that time is a critical resource in innovation), but often also reduces the cost. In the past Western companies especially have done that in Asia with manufacturing, and now they have begun the process with IT, back-office processing and pharmaceutical trials and R&D. Asian companies should also begin to tap into Asian capabilities to reduce their capital requirements. Furthermore, it is possible to strike mutually agreeable contracts with other companies who are also under competitive pressures, for example in terms of payment, part equity part upfront payment, revenue sharing and so on. In this environment of limited resources and increasing competition, Asian companies have an interesting opportunity to grow through partnerships. For this, it is important that they realize that a win–win mentality is a must.

CONCLUSION

Innovation requires resources. Without the appropriate human and financial resources creative entrepreneurs will not be able to achieve their ambition. The momentum of Asian innovators may well be stopped if they do not have the quantity and quality of technical infrastructure, engineering, design and management skills, creative employees and risk capital.

The availability of these resources depends on governments initiatives. We hope that some government leaders will heed the call to put in place the systems to develop these resources. But managers can also take some action to overcome the temporary shortages with which they will be confronted.

Engineering and design skills can be developed inside the organization, and managerial skills can be kept at a high level by maintaining good networks with your peers in industrialized countries. Financial needs can be covered by finding alternative sources inside or outside Asia and perhaps also by reducing the need for financing through creative partnerships with suppliers, customers or users.

Now that we have presented you with some inspiring examples of how to create an innovative culture, link better to the market and mobilize the necessary resources, we are ready for the next question: how do we make money out of our innovative efforts?

Notes

1 Hobday M., 1995, *Innovation in East Asia: The Challenge to Japan*, Edward Elgar, Cheltenham.
2 As reported by NASSCOM, Indian IT industry association, and quoted in *Financial Times*, 18 June, 2004.
3 For an example, see Innovators Struggle to Sell Their Ideas, *Financial Times*, 18 June, 2004.
4 Article by Takahiro Fujimoto, a professor at Tokyo University, quoted in (Still) Made in Japan, 7 April, 2004, *The Economist*. The original paper was published in Japanese in *Bungei Shunju* in November 2003, translated in *Japan Echo* in February 2004.
5 Hagel J., 2004, Offshoring Goes on the Offensive, *The McKinsey Quarterly*, (2): 82.
6 See the special coverage on Asian designer, 'Asia Life: Meet the Designers', *Far Eastern Economic Review*, 5 August, 2004, pp. 66–72.
7 Müller, U., reported in 'Rising Star', *AutoAsia*, October 2002, pp. 30–1.
8 Ghoshal S. and Piramal, G., 2000, *Managing Radical Change: What Indian Companies Must Do to Become World Class*, Viking, London.
9 Carney M. and Gedajlovic E., 2000, East Asian Financial Systems and the Transition from Investment-driven to Innovation-driven Economic Development, *International Journal of Innovation Management*, **4**(3).

7 Profit management

Hitherto, we have highlighted many challenges specific to innovation in Asia. In Chapter 5, we discussed the hurdles related to markets and marketing for innovative companies in Asia. And in Chapter 6, we elaborated on the scarcity of human and financial resources required to develop innovative products.

With this backdrop of challenges, it is clear that companies need to organize themselves to profit from their innovative offerings to justify the efforts and resources they mobilize in their development. Without appropriate profit streams from innovations, innovation initiatives cannot be sustained in companies.

Extracting profits from innovations in Asia is a different ball game from that in the industrialized countries. First, companies in Asia have to fight against illegal imitation, against which there is often little effective legal recourse. Second, often companies are confronted with fast imitation by competitors. Such imitation of products/services/business models is made trickier because it is often legal in nature. Imitation is obviously an international phenomenon, but it is a common strategy of Asian companies.

Both illegal imitation and fast legal copying present unique challenges to Asian companies. In this chapter, we highlight the extent of the problem in Asia and how it harms Asian consumers, governments and companies. We then discuss the root causes of the problem. This will lead to a better appreciation of the creative solutions to these problems as we have observed them through some of our case studies, and which may once again inspire you. We conclude the chapter with a summary of thoughts we believe will help companies in Asia to derive financial benefits from their innovative efforts.

ILLEGAL IMITATION

Counterfeiting is big business. The Counterfeiting Intelligence Bureau, part of the International Chamber of Commerce, believes that 5–7 per cent of global trade can be attributed to counterfeiting.[1] By OECD estimates, this is equal to about US$450 billion a year. This leads to significant losses by firms around the world, leading to job losses. However, due to the nature of counterfeiting, it is difficult to discover its true extent and exactly who is active in the trade.

One often reads about counterfeiting in the popular media. But contrary to what is often perceived, counterfeiting is not only about branded consumer goods. In fact, the more important counterfeiting is in medical drugs, creating health problems, or components for machinery, aviation and the automotive industry, thereby creating serious security and safety problems.

Asians: the unnoticed victims

Based on empirical evidence, it is commonly accepted that there is a problem with illegal imitation in Asia. Stories abound of firms from developed countries, especially Western multinationals, who are victims of the illegal imitation taking place in Asia. Which foreigner has not been approached on the streets of some Asian cities with fake watches? And Lacoste's famous crocodile has been on sale everywhere in Asia's backstreets and makeshift bazaars. It seems a bad business for foreign multinationals but a good business for the local budding entrepreneur. However, what is seldom seen is how much Asians themselves suffer because of the illegal imitation taking place in Asia. Asian consumers suffer because they cannot be sure whether they are getting the right products. Asian governments not only suffer a poor reputation in the international community but also lose out financially. Finally, Asian companies find it difficult to secure profits from their innovations.

The bulk of Asian consumers belong to the lower and lower middle class and are poor. They are also generally unaware of their consumer rights, where they exist. The lack of regulatory standards and ineffective implementation of the law seriously harms or even kills people when they consume difficult-to-identify counterfeit products. Counterfeit medicine, for instance, can be damaging to health and is sometimes even toxic. A WHO survey between January 1999 and October 2000 found that 60 per cent of fake medicine cases occurred in developing countries. In some instances flour has been put into capsules and passed off as medicine.[2] Medicines for HIV/AIDS, malaria and tuberculosis fall into such categories; the low prices of these fake medicines attract consumers. In mainland China, fake drugs can account for 40 per cent in some regions. And according to the Association for Asian Research, 192,000 Chinese died of fake drugs in 2001 alone.[3] In India, many people die or develop serious illnesses every year due to fake or poorly produced alcohol. Similarly, impure and fake milk caused the death of infants in China in April 2004. There have been a number of incidents of aeroplane crashes that are attributed to the use of fake components.[4]

Due to widespread counterfeiting, Asian governments also lose out. Foreign investors are cautious in sharing know-how and having local partners. Local companies become less interested in innovation, thus producing lower growth than possible for the economies. All of this means a loss of potential job creation and tax revenues. Since counterfeit goods are sold through clandestine channels, tax can't be collected. Counterfeiters often offer poor working conditions and also employ children, so governments become essentially ineffective in regulating workplaces. On top of all this, there is a huge reputational loss in the international community. In fact, that is one of the reasons why China claimed in 2004 that it is the biggest victim of pirated CDs.[5] The State Council, the country's Cabinet, estimated that China's counterfeiting crisis has cost the government at least US$19 billion in revenue annually.[6]

The illegal imitation of their products causes lost sales, profits and growth opportunities for innovative companies. Management time, itself an important and scarce resource, is taken up fighting legal battles and/or managing public relations. This is illustrated by the case of VazBuilt in the Philippines.

In 1991–92, VazBuilt, a developer of low and medium priced, top-grade houses using patented modular housing technology in the Philippines, was approached by a company called LPHI-Ayala, a subsidiary of the Ayala Group (one of the most prominent business groups in the Philippines). Thrilled with an opportunity to work with an Ayala company, the founder and CEO Edgardo Vazquez thought it was a good vehicle to achieve his mission of bringing top-quality, low-cost modular housing to his country. By the time Vazquez received the patent (filed earlier in 1990) from the Philippine Patent Office for the invention of the modular housing, his company had already built several hundred units for LPHI-Ayala. The partnership continued well even after the patent award. In the meantime, the World Intellectual Property Organization (WIPO) in Geneva gave VazBuilt the Golden Award for the Best Invention of the Year in 1995. Also in 1995, to speed up the process of building home units for LPHI-Ayala projects, LPHI-Ayala asked VazBuilt to cast its modular housing components on the construction site itself. Seeing the progress, in the middle of 1996, the Ayala Group proposed that Vazquez form a joint venture with it. The property market was booming; the Ayala Group wanted to make inroads into the middle and lower segments, diversifying its focus solely from high-end property development. Edgardo Vazquez wanted to see more commitment and details before striking the deal; finally, he rejected the relationship proposal. For the next two years, a series of LPHI-Ayala orders to VazBuilt were cancelled. In September 1997, VazBuilt officials found that LPHI-Ayala was building houses similar to VazBuilt's. After serving notices and trying to confirm its suspicions of patent infringement, VazBuilt launched a legal battle in 1999. The Ayala subsidiary claimed that its offering was based on other European patents. The battle continued at least until early 2004. We leave it to the courts to decide who is right or wrong in this case. The point is that legal battles are costly and may not always help the innovator, even if a court case is won. Considering the lifetime of patents, if and when VazBuilt wins, it will have no chance to extract rents from its patents; LPHI-Ayala is using its European licences till date.[7] Meanwhile, VazBuilt has had huge financial difficulties. Banks have recalled the loans that were extended during the relationship with LPHI-Ayala. Even finding lawyers capable of fighting the well-resourced and influential Ayala Group has been a tremendous challenge. Edgardo Vazquez has relied on the use of public media to gain sympathy and attention, but this took a lot of his and the management team's time.

There are also two recent lawsuits, one between Taiwan Semiconductor, the largest maker of custom chips in the world, and SMIC, the major semiconductor player from mainland China and the other between D-Link of Taiwan and Via, the world's second largest maker of chip sets for PCs based in Taiwan. Both are high-profile cases of industrial espionage, either directly or through poaching key employees. In both cases, in addition to the legal battles, there are huge media battles going on, which, needless to say, consume an enormous amount of management time.

At the moment, in Asia there are not strong brand products and consumer protection is weak. However, in the future when this may not be the case, it is not difficult to imagine that Asian companies will be swamped with warranty claims and other legal liabilities due to illegal imitation. The brand image may also be severely tarnished.

Causes and hurdles

If illegal imitation is not productive for Asian consumers or governments, and discourages Asian companies from being innovative, why does it thrive so well in Asia? In addition to the oft-cited weakness in the area of legal frameworks and enforcement, there are three other factors related to the characteristics of the market: the brand obsession of many Asian consumers, ethical issues and the socio-development agenda of governments, and supply side incentives.

Legal framework and enforcement

Most Asian companies and countries start from a low point in terms of appreciation of intellectual property (IP) protection. And, in many countries, the first initiatives to improve on this situation go back less than a decade. The laws are being developed under pressure from their commitments to the WTO and sometimes foreign investors. However, given the limited resources, a million developmental priorities competing for attention, and a lack of real experience in protecting IP rights, legal enforcement is also weak in Asia. This is not helped by the fact that most Asian governments do not even realize how important the protection of IP laws is in the scheme of things. There are only a few useful quantitative studies about the financial damage caused by ineffective IPR regimes, as the production and sale of counterfeit products is done on a secretive basis and there is no clear way to estimate. At best, Asian governments' response to these issues is patchy, depending on who is exerting pressure. For example, Southeast Asian governments have recently been trying to crack the fake CD market, mostly due to pressure from the US government. But fake handbags, shoes, spare parts – you name it – are freely available. Taiwan has passed new laws that provide greater protection to the high-tech industry but not for other sectors. The reason was clearly the pressure from an industry that is so vital to the economy of Taiwan.

Perhaps things would start changing drastically if foreign companies agreed to a mass boycott of a country. This seems unlikely in, say, China where FDI is increasing every year to unprecedented levels. Change may also occur when Asian countries themselves start producing a lot of own brands and thus intellectual properties which they would want to protect and benefit from. However, this is likely to be patchy too, depending on which industries are affected: in Taiwan it is the high-tech industry that has received favourable attention; in Thailand it may be Thai silk because of brand names like Jim Thomson. So, the lack of legal frameworks and enforcement are significant reasons for counterfeit products in Asia.

More than others, Singapore has lately been investing significant resources in IPR public education, public policy and law enforcement. The aim is not only to protect IPR, but also to encourage more local IPR registration. Similarly, at the regional level, Japan has been investing in efforts to lead a trilateral policy dialogue since 2001. Its aim is to develop better IPR regimes among Japan, South Korea and China. Japan considers this to be important, especially given the level of trade among the three countries. While it is too early to see any results, the three sides had agreed to set up an information exchange system to experiment with electronic document exchange and to jointly compile a technical dictionary related to IP. They also wanted to explore the possibility of sharing patent search and examination databases.

It is true that most Asian countries at different points in time have recently become members of various international treaties related to IP protection. However, the implementation has been mixed despite the mandatory requirements. Due to the different institutional systems in Asia, it is difficult to compare IP protection regimes objectively. A ranking of perceptions of IP protection in a series of Asian countries, as reported in *The Global Competitiveness Report* (2002–3) published by the World Economic Forum, is given in Table 7.1. The least one can say is that, with the exception of Singapore and Hong Kong, most Asian countries do not score as best in class.

Table 7.1 Ranking of perceptions of IP protection

Global ranking	Country	Score (on a scale of 1 to 7; higher score indicates more stringent protection)
12	Singapore	5.7
17	Hong Kong SAR	5.2
27	Taiwan	4.6
29	South Korea	4.5
33	Malaysia	4.4
37	Sri Lanka	4.0
38	Thailand	4.0
45	China	3.6
51	India	3.4
64	The Philippines	2.7
72	Indonesia	2.4

Source: *Global Competitiveness Report* 2002–3, World Economic Forum

Obsession with brands

On the demand side, Asians are generally obsessed with branded products from developed countries. This is especially true for goods like shoes and clothing that are visible to others when used. Asians are not interested in the most exclusive brands from the West, they are merely interested in brands like Nike for shoes and, say, Levis or Lee for jeans. These are simply the brands that people in Asian societies know well. With limited money at their disposal, counterfeits are a way to buy social respect – cheaply. (Contrary to the popular belief, it is often Westerners or Japanese visitors who tend to buy the exclusive brands of watches and clothes from Asia.) If functionality is important but difficult to replicate, Asian consumers are unlikely to buy fake products (for example refrigerators, ovens or food) intentionally unless they are conned into it.

Can a good legal system solve all the problems? We need look no further than the USA where online file swapping is widespread despite an advanced IPR regime. Although Western Europe is considered to have better IP protection laws than in the Asia-Pacific region, the economic impact from software piracy in Western Europe is not only higher than the impact from Asia, but is also the highest in the world.[8] Looking at Asia itself, Singapore's case is illustrative. Considered one of the Asian countries with a good IP policy (Table 7.1), software piracy is prevalent. Copyright infringement is a crime in Singapore, but according to a survey result quoted in the *New York Times*,[9] 75 per cent of Singaporeans had no personal objection to the use of pirated material. The same source, the Record Industry Association of Singapore, estimated that **500,000** people (out of a population of four million) download pirated music. The enforcement, however, has become stricter since the signing of a free trade agreement with the US in May 2003.

The laws might be a deterrent to piracy, but in many countries, the basic problem is that consumers see artists and production companies as rich already and well rewarded financially. Thus, consumers think that they are not really harming anybody – it is a mistaken case of a 'victimless crime'. Changing these attitudes in terms of piracy, especially of digital goods, is a huge challenge around the world.

Ethical issues and socio-development goals

Like all other countries, Asian countries want to develop. This need has to be balanced with the international IPR protection agreements with which these countries are in principle in agreement. High prices of patented goods that can affect the basic welfare of citizens make citizens uncomfortable, forcing governments to close one eye (or both). This is especially true of high-priced patented drugs from American and European companies. Countries like India have their own generic/cheaper versions of some important drugs for serious diseases like HIV/AIDS. These drugs are as effective as the alternative Western drugs, but are highly questionable under international patent laws. The rest of Asia can also get such low-priced drugs from India or other countries like Brazil that also produce such versions. These generic drugs may no longer be allowed when all WTO members have to comply with the TRIPS Agreement (Trade-Related Aspects of Intellectual Property Rights) by 2005. This is creating a huge ethical debate that is beyond the scope of this book. It is not only a question of the health of citizens but also the potentially huge negative effects on the productivity of Asian countries themselves.

Also worrying are the prices of information technology tools, which are important to economic growth and productivity. The high prices of standard products such as Microsoft Windows and Microsoft Office are disturbing to Asian governments from South Korea and China to Malaysia and Thailand. Thailand, China and Malaysia want to give their citizens a chance to access the tools at low price points and their domestic IT companies to grow. They are putting in efforts with Japanese companies to develop and promote their own version of the alternative operating system Linux. Malaysia and Thailand see IT as an opportunity for their citizens that can allow them to move up the economic ladder. So, these governments have promised low-cost desktop computers.

Asian companies know well that many countries in the world have achieved a high degree of economic prosperity through copying. This is even true for most of Europe, USA, Japan and even the Asian tigers. That makes it difficult for Asian developing companies to accept some of the conditions of the TRIPS Agreement, which they feel will only disadvantage them due to their low starting point. Furthermore, they feel disadvantaged because the old civilizations of Asia did not have a concept of patenting. For instance, herbal medicines have not been patented in China and India because the knowledge was just passed from one generation to the next by word of mouth. Now, some companies from the industrialized countries come and force their intellectual property rights on Asia and command high prices for what Asians consider they have known for generations. And companies and individuals in Asia often do not have the expertise and financial resources, or even awareness, to file for IP protection on a global basis.

Supply side incentives

Why do companies produce counterfeit goods? The lure of easy money, by addressing the market demands from Asian and other developing countries and the industrialized countries, is clearly a good incentive. This is encouraged by the lenient legal enforcement system. Even non-Asians come to invest in Asia in counterfeit production lines, subject to how much knowledge is required to produce counterfeits of a given sophistication.

However, the fundamental reason behind counterfeiting is that it is difficult to innovate in Asia. Given the challenges in the marketplace (as discussed in Chapter 5) and the lack of adequate resources (as discussed in Chapter 6), it is sometimes easier and more productive just to copy something that has already been proved successful.

What companies can do to solve these issues is the subject of discussion in the rest of the chapter.

STRATEGIES FOR CHANGE

Be tough and take legal action when necessary but not counterproductive

Only a few Asian companies are aware of their intellectual property rights, and, most Asian companies have few IP rights other than trademarks. Thus, it is not surprising that, more often than not, it is trademark infringement, which is perhaps also the easiest to discover, that Asian companies have to fight against. Not taking any action against an undesirable activity is often equivalent to encouraging it. Thus, it is worthwhile taking legal action against the infringers of intellectual property. VazBuilt of the Philippines, for instance, was built on its patent and found it necessary to file a lawsuit, as have many Taiwanese electronic companies.

International companies from developed countries, on the other hand, tend to be reluctant to take legal action unless they want to make a point about their seriousness. They often find such legal actions ineffective, and prefer to lobby their home governments and the country of operations for better legal frameworks and enforcement. A survey by JETRO in Thailand found that more than 70 per cent of Japanese companies do not use the effective TRIPS provision to fight counterfeits by enlisting help from the customs department of the country, even when they think that counterfeit products in a country are imported to some extent.[10]

But before a company in Asia takes legal action against illegal imitation, it should carefully think about the quality of legal infrastructure in the country. While Taiwanese electronics companies may feel comfortable doing so, a construction company like VazBuilt in the Philippines has a relatively lower chance of success. In addition, a company should make a judgement as to the lengths it is going to go to win the case. More often than not, companies have no idea or proper quantitative estimates of the incurred or potential damage. Legal battles cost a lot of money and are time-consuming, but most companies underestimate this and are often driven by their emotions.

Lawsuits may also have a negative effect on companies. Immediately after VazBuilt filed a lawsuit against LPHI-Ayala in late 1999, the company not only lost all its contracts with Ayala but all its investments became unproductive. The banks recalled their loans, threatening VazBuilt with bankruptcy, which was averted only with family money. Often, new clients avoid companies caught up in legal battles, and if there is an apparently more powerful party in the lawsuit, they tend to prefer the bigger and stronger party.

Finally, companies should explore solving matters amicably because legal battles are too long and the rewards are too small due to the value assigned to things like intellectual property. Sometimes the alleged infringer should be given an opportunity to save its public face. Cisco and Huawei's case is illustrative in this situation. Cisco claimed that Huawei, the Chinese telecommunication network equipment manufacturer, had copied its source code illegally, leading to a high degree of similarity between Cisco's and Huawei's routers and switches. Given that the errors present in the source code of Cisco were present in that of Huawei too, it is quite likely that Huawei infringed the IP rights of Cisco. After initially going public with such information and filing lawsuits against Huawei, Cisco quietly withdrew its lawsuit and came up with a deal, the details of which are not available to the public. Given that it is important in China to maintain one's public face, Huawei (one of the largest companies in China) was offered a chance to maintain its public image both in China and overseas.

Learn from piracy: turn a threat into an opportunity

Sometimes piracy is an indicator of how customers think about a certain offering. And, therefore, there is much to learn from piracy. Generally, companies initially reject illegal imitations. The prices of counterfeit products are almost always lower than those of the originals. So, the original product companies tend to look down on the quality of the products. Low-quality, illegal imitation is supposedly targeted at the market segments which are not their focus, giving them a feeling of security. But such an outright rejection of the situation may be flawed. Imitation offers a company an opportunity to think about that low-end market which wants the product but cannot afford the original. Creative Technologies of Singapore realized that the illegal imitations of its famous soundblaster cards were available cheaply and sold like hot cakes in China's booming PC market. The company took the unconventional step of using its facilities to manufacture unbranded cards and selling them in the Chinese market to gain a slice of the market that it would not otherwise have achieved with its high-price, branded product. Thus, it did not engage in a large-scale legal battle with small vendors across China, which would have been unproductive, but it gained some more market share.

Alternative brands can also be created by companies to target the low-end market. For that reason, the luxury chain Banyan Tree Hotels and Resorts now has other lower end brands like Angsana to cater to the middle segment of the market.

However, when the illegal imitation is of good and comparable quality and existing and new customers start buying the counterfeit goods, a company tends to find solace in legal protection. But, the chances of success are limited and returns are often not that significant in Asia. Companies also tend to overlook the signals from the market, which make it clear that companies need to innovate – either in their offering or business model – to create more value for customers. If, however, it is a high-quality imitation, it could be argued that the original was not a sustainable proposition. If companies are vigilant and focus on their customers, they can try to do a better job at tackling piracy by reinventing their offering.

It could be said that Hollywood's strategy of releasing DVDs a few months after the release of films in cinemas is flawed and unsustainable, because DVDs of the latest films are freely available in many parts of the world including in Asia as soon as – and frequently even before – a film has been released in the US. Similarly, the strategy of record companies to bundle good and bad songs on one CD to sell more is disliked by customers. No wonder they download only the songs they like from the internet. In both cases, the suppliers (be it the film studios or the music editors) attempt to impose a business model on the customer, which the customer refuses to follow and so looks for alternatives. In these cases, a company is better off by innovating through a better understanding of why the customer does not adhere to its business model.

It is said that customers want everything free. This, we believe, is not true. If you were to ask the people around you how much they are willing to pay for a music CD, we are sure that, with the exception of a few, most people will give a number bigger than zero. Apple decided to give customers their freedom through Apple's iTunes Music Store. In fact, at US$0.99, they pay roughly the same rate per song as when they bought CDs. But now they can buy only one song, if they like. In the reverse auction system of Priceline.com, nobody asks for an air ticket for zero price. Customers try to get the best deal possible, but what they offer to pay is generally what they consider legitimate.

Annalakshmi, a successful Indian vegetarian restaurant chain, has a unique concept: 'Eat what you like, pay what you like'. Located in four countries (India, Malaysia, Singapore and Australia), the restaurants are lavishly decorated with carvings, paintings, statues and pillars from India and the menu is elaborate. The staff is largely voluntary – housewives with the appropriate culinary skills and doctors, engineers, managers, bankers and others to serve the customers. Annalakshmi helps to sustain the artistic and charitable activities of the The Temple of Fine Arts, an organization which has the objective of spreading Indian culture; the organization teaches dances, holds cultural festivals and offers free medical services through its doctor members. Despite the payment policy of the restaurant in Singapore, the people who eat there almost always pay at least what they would normally pay in a good restaurant for a similar meal. This example shows that customers are usually prepared to pay what they consider to be legitimate.

Thus, the important thing is to create value for customers in a way that *legitimizes* the legal rights of the company in the eyes of customers. Examples of this are easy to find. The publishing industry offers different prices of books in different countries. The Western books, for instance, are priced so they are cheap enough in Asia to be affordable and yet more expensive than photocopying. Similarly, a number of countries around the world are putting price ceilings on a number of locally produced products in a belief that this will reduce piracy. For example in Malaysia, local artists' CDs are RM21 (US$5.53) and foreign artists' CDs are RM29 (US$7.63).

Imagine that illegal imitation is 'legal' and try to fight it competitively

While a person from a developed country will more often than not fall sick after drinking tap water in most parts of Asia, the native people in those countries drink the same water without falling sick. Asians cannot wait for the water and air to get cleaner *before* they start living. Unhygienic environments create stronger immune systems.

Similarly, the companies in Asia should, we suggest, develop the agility to live in their world of poor IPR regimes. This would allow them to become more resistant to the competitive threats arising from illegal imitation. Without relying on IPR regimes, the innovative companies of Asia should focus on finding ways that offer them the ability to extract profits from their investments in innovation. And, with such a frame of mind and capabilities, they will be fit enough to fight the competition from anywhere in the world.

The best companies are not built on IPR alone. Think about the coffee chain Starbucks – theoretically anybody could compete with it by opening similar stores. Starbucks did outrun its imitators easily. It is often possible to work around the patents to arrive at the same result in a different way. So, it is important to build competitive advantages that help to fight imitation – legal or illegal.

Asian companies should start thinking of all imitation, illegal or legal, as if it were legal and start finding ways to thrive. Such an attitude is especially important for business innovations. Legal imitators are perhaps more dangerous than illegal ones. Such imitators can also incorporate the lessons gleaned from the mistakes of the innovator.

The following are some of the ways that innovative companies have utilized to build a sustainable competitive advantage over legal and illegal imitators.

Business and process secrets

In general, the simple copying of products or services is not effective. Even if the product or service is copied well, the copycat company may not be as effective because there are many invisible sources of success hidden in the processes of bringing the offering to market in an economically successful manner. So, the companies should put in their efforts, resources and procedures in protecting those secrets.

Lapid Foods offers chicharons, one of the most common snacks in the Philippines. But, the product is still special compared to its copycats. Despite Lapid talking about its 'freshly popped chicharons' and even discussing the recipe openly in public, no one is able to offer as good a chicharon. The reason: the recipe and temperature control took one year to perfect. While a competitor may know the recipe, it is the secret tacit knowledge that makes Lapid's chicharons perhaps the best in the Philippines.

The same idea is exhibited on a much larger scale by Li & Fung of Hong Kong.[11] Evolving from a sourcing agent, today Li & Fung is a successful global supply chain manager. While supply chain management may indicate complexity but not novelty, behind Li & Fung's successful transformation are many innovations and deep knowledge that makes its successful model difficult to copy. Reflecting its concept of borderless manufacturing, Li & Fung buys from around 7500 suppliers and manufacturers in some 40 countries that possess specialized product and manufacturing capabilities. But, it does not own any of the facilities involved in processing the raw materials into finished products, or any of the equipment that transforms the semi-finished or finished products through the stages of production. To cope with the complex challenge of managing suppliers and manufacturers and the flow of parts and materials along the production chain, Li & Fung provides intellectual and technical leadership and incentives that hold all the suppliers and manufacturers in the network. It manages the information flows in the entire network, as well as the relationships with customers and suppliers, to achieve reductions in cycle time, cost and risk. At the same time, it uses IT to manage the relationships and communication among suppliers and customers spread around the globe. All this requires organizational innovation to manage its complex network. Li & Fung cannot and doesn't run this business on the legal protection of IPR. Its strength lies in its deep knowledge, immense network and a human and technical infrastructure that allows rapid exploitation of generally available knowledge. This is difficult to copy.

Napoleon L. Nazareno, CEO of Smart Communications in the Philippines, would argue that one of the key reasons that Smart is highly profitable despite heavy competition and copying of its business model is because of its sophisticated back-end infrastructure. Most of its approximately 12 million customers are subscribers to a prepaid service in the Philippines. When a call is made using the credits on the prepaid card, Smart deduct the appropriate amount from the card. Since credit card penetration is low in the Philippines, prepaid cards are not backed by credit cards. Thus, when Smart's backroom deduct the balance from what is left as credit on the prepaid card, it has to do so real time. Otherwise it would have a lot of losses, if calls could be made after the prepaid credit is exhausted. Developing and maintaining an infrastructure that can react in real time is something that is not a visible challenge, but this is where Smart invested and built its competitive advantage.

How does one protect the company's sensitive secrets? The weakest links for knowledge transfer between organizations are the employees, be they unhappy employees (current or former), contractors or affiliates. Taiwan Semiconductor has been suing SMIC, because it thinks that SMIC poached its employees and later forced them to divulge some secrets. The case between Via and D-Link is one where the latter accuses the former of corporate espionage. Such lawsuits will take their own time to resolve. But how does one prevent the leak of secrets from corporations?

Lapid Foods treats its 100 employees like a family. The company gives them incentives to stay, provides them with accommodation, thus creating a moral bond. Such a policy may work well in a small organization.

But the challenges for large companies are different. And, given that there is plenty of information – storage devices, miniature electronics gadgets and free access to the internet, it is difficult for companies to do much. Some companies have started keeping a copy of all the emails sent out on a central server, some do automatic checks for sensitive information being sent out, some focus on physical checks. In July 2003, Samsung banned the use of camera-phones in some of its factories, for fear of spies snapping pictures of new models.[12]

These are some things that companies have been able to do, but when an employee leaves, he/she also carries information in his/her head and that is the most difficult to identify. Often, such people leave with client lists. One expert suggests inserting a few fake contacts into a client list, with email addresses that circle back to you.[13]

Shin Satellite, which has developed a ground-breaking technology for IP communication over satellite, has found another interesting way. There are very few people in the company who have access to where other parts of technology are being worked out. Other such centres are spread around the world, and the company is protective of which institutions or individuals are involved in the technology development. The diffusion of knowledge over many different sites and people makes the task of copying difficult.

Brands and experiences

Innovative companies tend to think too much in terms of their offerings and how they can satisfy their customers' *functional* needs. For profitability, it is important for them to build an emotional appeal too – one that lasts and continues to attract customers even when competing offerings are available. Such emotional appeal could be built in a variety of ways.

iRiver found an interesting way to address the challenges. It sells hardware that is often copied, albeit imperfectly. One of the main reasons that fake iRiver gadgets are unable to enter the mainstream is because of the free software upgrades which are available only for original iRiver products. The customers engage in online communication with the iRiver community around the world to exchange thoughts and experiences. They could also make suggestions to iRiver or criticize it, and all of this is open and transparent to other people. Now, iRiver is creating new opportunities to enhance the experiences of its customers. It is opening its first physical service stores in South Korea. But this idea is more than just aftersales service which can be easily obtained through partners or online. As the name – iLounge – of these stores reflects, iRiver is basically creating 'a new community place to hang out and share experience and joy!' The positioning of the iRiver brand as lifestyle products instead of merely functional ones continues to give iRiver a special appeal. As a consequence, iRiver has more than 20 per cent of market share in the US and more than 50 per cent in South Korea.

Banyan Tree Hotels and Resorts offers an emotional appeal in another way, as reflected in its tagline 'Sanctuary for senses'. 'We wanted to create something that would conjure up certain images – romance, rejuvenation, intimacy – in people's minds', said its CEO Ho Kwon Ping.[14] 'Everything we've done has tried to reinforce that.' The concept is unique in Asia and had its own risks, but has a clearly articulated objective of creating a unique experience. The soft approach distinguishes the Banyan Tree resorts from its competitors, allows Banyan to charge a premium and connect with customers on an emotional level.[15] The company could raise prices even during the Asian financial crisis. Recently, it spun off popular features of its resorts – spas and retail shops – to leverage the brand. The spas boast tropical serenity and lately have been put in as branded spa outlets in several countries across Asia and Africa, and 'draw on local inspiration to blend seamlessly with the natural environment'. The company regularly wins international accolades and is considered to be special and innovative by its customers. The special experience has created an emotional appeal and loyalty that other resorts may find hard to replicate and steal.

Food can also sell by its emotional appeal. Let us look at the example of Lapid Foods again. There are only a few Lapid outlets, which not only helps to fight counterfeit products (Lapid is a very common family name in the Philippines), but also creates a conscious temptation for a product that is based on impulse buying. Now, the whole experience of going to Lapid Foods is 'going there to do something special'. The snack shop even provides valet parking and umbrellas in the rain for those who come to buy chicharons. That is how one creates a special relationship with customers. The company has never advertised and relies solely on word-of-mouth publicity.

Tiger Motors of Thailand introduced a new model of its Nano electric motorbike in conjunction with the new Warner Brothers' movie, *Torque*. Riding on the atmosphere created by *Torque* in which motorcycles play a central role, Tiger introduced only 100 new Nano Torque Limited Edition motorbikes. The Nano Torque Limited Editions have more power and a higher price tag (Bt24,900) compared to Bt19,900 for an ordinary Nano bike.[16]

Patkol has created and sustained demand by creating opportunities for experiencing its product. To help potential consumers to experience the benefits of tube ice, Patkol started giving out clean tube ice for marriages in Thailand. When block ice had been used in Thailand, hygiene was a big problem. When the guests used the clean ice for marriages, they liked it. This had a knock-on effect, as the CEO of Patkol explained: 'when the guy next door wanted to arrange a party, he couldn't arrange it without sanitary ice made using Patkol's machines by Patkol's customers. In fact, we heard of one case where the party organizer had to drive 30km to get ice for his party.' Indeed, Patkol started urging its customers to put in place a better distribution infrastructure. This shows how, with its simple technique, Patkol's tube ice became more popular than ever, benefiting the company.

The point is simple: emotional appeal is essential to make innovative products create economic value for the producers.

Market dominance

One way to beat rapid legal imitation is to have market dominance, or at least aim for it as soon as possible. Markets where there is a dominant supplier with a big market share are unattractive to competitors, even those who copy: the market share they can obtain is often too small to have a profitable operation and write off their fixed costs.

In the early 1980s, it became apparent that IT would have a bright future and the existing Indian IT professionals were in demand. As there were not enough IT education providers at that time, two entrepreneurs started a new private company called NIIT in India. As the demand exploded, the company hit upon the concept of franchising, which was perhaps the first to be applied to education. Today, the company is a very well-known brand throughout India; education from NIIT is often highlighted even in matrimonial advertisements as it normally leads to a good career. NIIT has a market share of 40 per cent, with an alumni base of more than 1 million students.[17] A large number of employees have been former students too. Using the pool of students, the company then ventured into customized business applications which has also done very well. Having gained a reputation at home, NIIT now has franchises in more than 30 countries and expects to have more than 150 franchises in China within a few years. It has entered the e-learning space. With its strong focus on quality expansion, NIIT is number one in India in IT education and among the top 20 global IT training companies. NIIT's network of franchises, large number of training courses (electronic and personally delivered), strong brand image and high degree of localization, makes it difficult for new players to penetrate its markets.

For new players to create market dominance is not easy. There needs to be something to ride on. Reliance Infocomm handed out heavily discounted mobile phones to the shareholders of its parent Reliance Group, which had the largest number of shareholders in India. As shareholders got such good deals, they tended to take subscriptions which were also given on special terms. Riding on the shareholder base in this way, Reliance Infocomm helped itself to become the market leader in its field in less than a year.

Similarly, domination in the market can also create a huge competitive advantage in attracting talent. For instance, because of its image Samsung Electronics is able to attract some of the most talented people in South Korea to work for it, creating an advantage that is difficult for its smaller competitors to overcome.

Business alliances

If market domination is not possible, a wide set of alliances create advantages that are difficult to replicate. For instance, Li & Fung's network of 7500 quality-conscious, cost-effective producers who deliver to a short deadline for its customers helps to keep its competitors at bay.

To fight the loss of 30–40 per cent of revenue from film piracy, South Korean film makers are looking at alternative sources of revenue. They are resigned to the fact that court action is having little success, and have responded by seeking alternative ways to make money. They are forming alliances with Hollywood film makers and are offering the remake rights to its best movies.

There are other kinds of alliances – social alliances – that can help companies in defending their market share. Realizing the socioeconomic development goals of low-income Southeast Asian countries, Microsoft, for instance, has taken the step of providing a low-price, Malay language version of Windows XP bundled with some new desktop computers. This was sold under the Gemilang project of the Malaysian government, which aimed to raise PC penetration rates in Malaysia from about 15 per cent currently to 35 per cent over a period of two years. In this way, Microsoft was able to fight off some of the social pressures from the government. At the same time, it was able to gain more market share and not lose everything to the low-cost Linux desktops. The bundling of products, as well as providing an offer in the local language, satisfied customers and quelled Microsoft's fears of export due to price arbitrage.

Through the social alliance with their stakeholders, Neowiz has gained social legitimacy in a different kind of situation. A leader and creator of the digital avatar industry, Neowiz is extremely popular in South Korea. With each avatar sold at US$4–5, some kids started spending up to W200,000 (US$150) per month on these avatars. These kids, mostly teenagers, don't own credit cards, so they use the fixed line phone to purchase the avatars, and the fee is reflected later in the phone bill. With such high avatar bills and little control in the beginning, parents were upset and complained to the government and threatened legal action against Neowiz. Of course, government had no laws against this. But Neowiz engaged with parents and the government, and offered some help in thinking about the issue. A cap of about US$30 per month was agreed, which the South Korean government then used later to apply to all other companies in the same industry. Thus, using its own initiative to become socially legitimate, Neowiz survived and prospered.

Speed in innovation

Rapid innovation is an effective technique used by many successful innovative companies in Asia. There are three effects of such a strategy when executed well: it creates a bigger mind share for the company, which often leads to bigger market share; imitators find it difficult to keep up with the pace of rapid innovation; and the resulting success keeps the organizational motivation high, reinforcing the momentum for rapid innovation.

In 2003, while all other South Korean and international MP3 player companies announced only two new products, iRiver launched eight different products during the same time and 13 new models, including video products, within the first half of 2004. This momentum created significant market talk and reviews about iRiver products, attracting more attention from the customers. Imitators usually have limited resources and are unable to offer so many new products. Often, the imitators wait for the product to be successful and such a profusion of products is confusing for them. iRiver was fully aware that not all products would sell well, but that had been incorporated into the planning. The products were designed in such a way that R&D efforts could be utilized across several models.

Smart Communications of the Philippines uses the same strategy to retain its market leadership. It introduces one new service per week as compared to its competitors, who at best introduce one new service per month. Smart is able to introduce novel services so fast because it nurtures over a hundred independent content providers, which are usually start up companies run by young, creative and enthusiastic people.

How does one create such rapid innovations? In Chapter 2, we alluded to some of the actions that can be taken. Part of the answer lies in creating a set of options or opportunities through rapid prototyping and testing when there is no crisis. For that, it is also important to create the right organizational culture and structures.

Smart Communications has a pool of funds set aside to give a small amount of money to help start-ups to develop some content for their users. The fund required is small as most of the infrastructure is shared with similar companies and controlled by Smart. Diverse companies come up with creative services. This is in sharp contrast to its close competitor Globe Telecom, which manages its own IT company that is expected to come up with innovative services. Encumbered with more bureaucracy and perhaps a less than perfect incentive structure, the rate of innovation at Globe was one-fourth that of Smart. Not only that, Smart marketing is lively and the organization is accepting of failure as it understands the culture of experimentation.

It is a similar story with Neowiz. At any given moment, it is experimenting with several new ideas. Since its wildly successful avatar business idea came out of rapid prototyping and testing with customers over the internet, the organization is always willing to allocate the small funds needed to try out new services. Having tried and failed counts more for an employee's success in the company as compared to not trying at all. At the same time, Neowiz is strict with the criteria of not letting unsuccessful experiments continue. Even when it entered the web card games business late, rapid prototyping and testing with customers allowed it to create experiences for customers that were more exciting than those offered by the creators of the industry segment. Now its main source of revenue is these games and not the avatars.

The National Library Board of Singapore has developed all its innovative services through the process of rapid prototyping. Staff members, based on their observations and input from the customers, propose new services for development. The services are quickly developed for testing in one of the branches where they are observed for a given period of time. If the success criteria are met, the service is rolled out rapidly in all the libraries across Singapore, incorporating the new learnings from the test offering.

Finally, you can go for markets that stimulate you to go quickly and then use what you learn in other market segments. ChoLam of Thailand first developed a business in adult clothing. But it realized that adults in Thailand didn't buy new clothes that often. So, it came upon the idea of designing clothes for young children. Children grow quickly and thus need new clothes more often. So, ChoLam learned how to produce its designs more quickly. And it is attempting to diffuse this approach in other parts of the business.

CONCLUSION

The basic premise of this chapter is that once you have mobilized the resources to innovate and you know what your customer needs, you also need to invest in profit management. You may remember from Chapter 2 that we consider an innovation to be successful only if we have an economic success, that is, a return that includes a reward for the risk that innovation entails.

Gaining these returns requires careful profit management. In Asia entrepreneurs' profits often disappear because of copying. This chapter has shown that illegal imitation is rampant in Asia and is harmful for Asian consumers, governments and innovative companies. While we would like to make a plea to Asian governments to invest more in intellectual property protection, we also realize that there are many reasons why this will take some time to be implemented. Therefore companies have to engage in strategies to cope with imitation.

Pursuing the enforcement of one's legal rights can be one way. But this often takes long and costly efforts and it is not clear whether the owner of the IPR can achieve a worthwhile return out of it. Perhaps the approach should be to try to obtain enforcement of legal rights, but combine this with a more managerial approach. The two major suggestions for inspiration that we gave here were: learn from what your imitators do and consider that all imitation is legal and that IPR is a windfall profit and thus you have to defend yourself with other competitive means. Indeed, there are examples of companies in Asia that have pursued these two approaches and we hope that this will inspire you.

Notes

1 See http://www.iccwbo.org/ccs/menu_cib_bureau.asp, International Chamber of Commerce Counterfeiting Intelligence Bureau.
2 Global Rise in Use of Fake Drugs, *BBC News*, 11 November 2003.
3 Fake medicine becomes rampant in China, Association for Asian Research, 26 July, 2003, www.asianresearch.com.
4 Vithlami H., 1998, The economic impact of counterfeiting, OECD report, www.oecd.org.
5 China claims it is largest victim of CD piracy, Agence France-Presse, 14 April 2004.
6 Milk tragedy shows that fake goods cost lives, Chi-chu T., *Straits Times*, Singapore, 23 April 2004.
7 This case has been covered widely by the media in the Philippines.
8 Piracy in Asia costs software makers $7.5 billion, *International Herald Tribune*, 7 July 2004.
9 Piracy is rampant globally, Landler M., *The New York Times*, 29 September 2003.
10 The Result of Survey on Damages Caused by Counterfeits to Japanese Companies in Thailand, May 2003, JETRO Bangkok Center, Japan Patent Office.
11 Mar P., Richter F.J., De Meyer A. and Williamson, P., 2005, *Global Future: The Next Challenge for Asian Business*, Wiley, New York.
12 Fong M., 2004, The Enemy Within, *Far Eastern Economic Review*, 22 April.
13 Fong M., ibid.
14 Slater J., 2000, Putting Down Roots – Banyan Tree Built a Name for Itself from Scratch, *Far Eastern Economic Review*, 25 May.
15 Slater J., ibid.
16 Tiger Gets into the Movies, 2004, *The Nation*, 9 February.
17 Das G., *India Unbound*, Profile Books, London.

8 Overcoming the underdog mentality

Up to this point, we have covered a few topics that are generic to innovation but which require special attention in Asia. We provided examples of organizational change. We discussed some specific ideas related to markets and the marketing of Asian innovations. We gave inspirational case studies on how to mobilize the resources needed for innovation. And in Chapter 7 we examined how to make a profit from innovation efforts.

In this chapter, we extend the analysis of the management of innovation in Asia into a topic that is seldom discussed in management, especially in the context of innovation. During the first phase of our research, to our surprise we discovered that firms in developing Asia were often reluctant to innovate, even when it was economically sound to do so, because they suffer from an 'underdog mentality'. This underdog mentality is created and reinforced by poor customer perception, negative media coverage and a lack of influence in the international bodies that set standards.

From the typology of Asian innovators described in Chapter 3, it is clear that not all companies suffer from this underdog mentality – it is not a hurdle for everybody. The group of innovation starters does not see this as its first hurdle to overcome. But the tradition fighters see it as one of the most important brakes on their development.

What do we mean by this 'underdog mentality'? In many of our interviews, we perceived that company executives thought that their firms did not have sufficient capabilities to innovate. In some cases, this was reinforced by a negative perception that the market had about the company's products. This attitude leads to a vicious cycle: managers of those underdog firms feel more incapable of producing innovative goods and services and being successful with them. And the lack of substantial innovation success resulting from this reinforces the feeling that the company does not have the appropriate capabilities to innovate. Here we are not suggesting that there is a lack of competence, it is more a *lack of confidence*. With such an underdog mentality, it is difficult to become a successful innovator. The vicious cycle becomes a self-fulfilling prophecy.

Clearly, not all firms in developing Asia have these problems, and some who have had these problems have overcome them in some insightful ways. As in the other chapters, we discuss these problems and try to provide inspiring examples of companies who have countered them.

THE PERCEPTION IN INDUSTRIALIZED COUNTRIES THAT ASIAN PRODUCTS ARE POOR QUALITY

In Chapter 5, we discussed the fact that markets in industrialized countries are on average more sophisticated than many Asian markets and hence understanding these markets is important for the aspiring innovator. We said that Asian companies may not understand customers from these markets due to geographical and cultural distances. Even when they do, many of these innovators continue to face the perception of poor quality goods and services by potential customers in those countries. Three quotes may illustrate what we observed during the interviews. Narayana Murthy, founder, chairman and chief mentor of Infosys, one of the top Indian IT service firms with a market capital of more than US$1 billion in fiscal year 2004, is reported to have said:

> In the early 90s, when we went to the USA to sell our services, most chief information officers didn't believe that an Indian company could build the large applications they needed. The CIOs were very nice to us, of course. They offered us coffee or tea, listened to what we had to say and then said, 'Look, don't call us – we'll call you.' We realized that there was a huge gap between, on the one hand, how prospective Western clients perceived Indian companies and, on the other, our own perception of our strengths.[1]

Yeap Swee Chuan, the CEO of Aapico, a company we have often used as an example in previous chapters, had a similar experience:

> I went to Germany to make a presentation to a very senior executive of a German auto company and several of his subordinates. After I made my pitch, the first thing he said was 'I have been to your country Thailand. Very nice elephants.' I immediately realized how he saw us and our capabilities.[2]

And in our discussions with ChoLam's management, we heard:

> At the very first, when we were launching our product in the market like in Saudi Arabia, they said, ah, the product looks nice. But they see the tag made in Thailand and they say that it should be cheaper than this. Their immediate reaction was one about the price. Now they have been buying for more than five years. And they had to try to communicate to their customers and educate them that this is a nice product with good quality, with good fabrication, with good design. It doesn't matter that it comes

from Thailand. It is possible to make a good product anywhere. Right now I believe that our products have already been established in that market because they know us well and they buy a lot from us.

In 2005, several years after the incidents referred to above, both Infosys (and the Indian IT industry as a whole) and Aapico are well known and well regarded in their fields of endeavour. Not only that, Narayana Murthy of Infosys is recognized as a global business icon. But the firms and the individuals had to fight the negative perceptions of the developed markets in order to establish their credibility and make their value proposition acceptable.

Of course, this poor perception is not unfounded. An important part of the reason is the quality, or rather the lack of it. For a long time, the quality of the goods produced in Asia has been poor and, let us be honest, it still is in many cases. It is clear, however, that quality is not an insurmountable problem. Many firms throughout the region have successfully implemented TQM techniques. But the doors that impressions of bad quality closed in the past are sometimes difficult to reopen. Furthermore, the design and packaging of Asian goods is often shoddy – they are functional in nature and have limited aesthetic value. Firms in Asia allocate insufficient resources (partly because they don't have them) to adapt the marketing message or the product itself to the market. Markets in industrialized countries are highly diverse – the US is vastly different from Europe, which itself is not at all homogeneous. And both of them are quite distinct from Japan. This poses significant challenges for Asian firms, especially when they do not have sophisticated techniques to influence consumer behaviour.

But, there are other, less rational reasons for the poor image problem. Low wages in Asia are seen as a reflection of low value added – without due regard for the concept of comparative economics and purchasing power parities. Indian companies are seen to do low-end work, whether in the IT or pharmaceutical world, because of the lower wages paid in India. China is widely seen as a low-end manufacturing centre. We know that this is no longer always right. Some high-quality scientific talent is employed and some leading R&D is done in India and China. While this may be still the early stages of their scientific efforts, companies and universities are increasingly investing in these areas.

Sometimes sophisticated consumers in the industrialized markets do appreciate Asian goods but in an odd way. It is not uncommon to see them wearing some ornaments from Thailand, necklaces or other accessories from India, having some antiques from China and face masks from Indonesia and South Korea. But that is it – Asian products are too often seen as cute and exotic, not as products that can be used by mainstream people in a mainstream functional way in society. Often, one hears references like 'So cute and also soooo cheap'.

Having said all this, many companies have been able to overcome these perceptions. Here are some interesting examples. In Germany, when Yeap

Swee Chuan, the Aapico CEO, was told about the elephants in Thailand after his extensive presentation of the technical capabilities of his company, he left the German company's premises knowing he would not get the contract. But, he has done a few things since then that have boosted the perception of his company's value among the top car makers in the world. First, he made sure that quality could not be seen as an issue. He worked hard to obtain international quality certifications like ISO and QS9000 and a sound balance sheet (he proudly says it is much better than that of Ford). But, he found that most executives who made decisions about whether he would get a contract had never been to Asia or had only been on short tourist or business visits. So, as a rule, he started inviting these executives to his factory premises. He did that with the help of his contacts in the local offices of those multinational corporations or the contacts with senior executives that he had directly developed by attending the right automotive conferences around the world. And, he remarked that 'the perception of us was poor overall until the senior executives visited us'. Further to that, he offered to let the visiting executives do a test and let the results speak for themselves.

Yeap regularly invites senior executives from car makers to his factories to show that his company is doing its best and should not be labelled as bad simply because it is Thai. For instance, when he had a problem once with Toyota, he apologized and invited the senior executives to the premises to examine the production and give him their advice. When the Toyota executives came, they were surprised to see how well organized his company was. They were pleased to see that it had made attempts to implement and enhance the Toyota Production System. Today, Aapico is one of the few Asian companies to be a tier-one supplier to most of its auto maker world customers. Aapico has also partnered with DaimlerChrysler's Lean Manufacturing Centre to advise other companies in the region, and has set up a joint venture with Siemens VDO.

To remove quality concerns, companies like the now well-recognized Haier, China's leading white goods producer, had to go through a similar tough route. Even after it had obtained the German authorities' approval, German dealers were hesitant about the quality of refrigerators from China. Therefore, Haier took off the trademarks from its refrigerators and those of German-made products and mixed the items up for the German dealers to try to distinguish the 'bad' from the 'good'. When the dealers failed to tell the difference, Haier got its first purchase orders and started building its brand equity. Soon after its market entry, Haier products came out top in the German government inspection departments of the quality of refrigerator products, including home-made and imported ones![3]

In the case of the Indian IT industry, the industry association – the National Association of Software and Service Companies (NASSCOM) – supported by the government tried to spread awareness in the key developed markets to explain both the quality and price value proposition. Today, India is clearly recognized as influential with its proven off-shore delivery model. As an indicator of quality, India had about 60 out of 80 SEI CMM[4] level 5 IT companies worldwide in early 2004. Similarly, many Indian IT companies have ISO and other quality standards certification.

Government efforts are especially helpful for young companies. For instance, the Seoul Industry Promotion Foundation helps SMEs in the Seoul area to achieve international quality standard certifications. Similarly, other projects like the Hi Seoul programmes help the SMEs in their branding efforts.

The Samtel Group of India had a different problem. It was trying to jump into the league of R&D outsourcing for Western companies but was facing huge perception problems. So, it bought a financially distressed German company to be the face of the R&D outsourcing business but actual R&D continued in India. The German unit also allowed Samtel to overcome the unspoken market resistance to its high-end cathode ray tubes for military, medical and industrial applications.

In Chapter 5, we also saw how the fragrance Lolita Lempicka, although produced by the South Korean cosmetics producer AmorePacific dissociated itself from its Korean origins to create a public perception that it is French and hence worthy of consideration by sophisticated consumers.

These examples and quotes show that the negative perception of Asian products in the sophisticated markets of the industrialized countries is a real issue. It will not go away by simply developing capabilities in making better and more innovative products and services. Firms will also need to change the perception to be effective. Our examples show that this requires a careful and well-planned educational process. The combined efforts of firms with government support may help the industry as a whole (and that is important), but Asian firms themselves will also need to proactively manage this issue of poor image in sophisticated markets in the developed countries. The challenge may become less important in the coming years, because of the general improvement in the image of Asian firms, but it will be a slow process. Successful innovators will have to take their own destiny in their hands and have the guts to go and convince the sophisticated markets of their capabilities.

EVEN ASIANS PERCEIVE LOCAL PRODUCTS TO BE INFERIOR

It is not only customers from developed countries who have a negative perception of Asian goods, but, very often, even Asian consumers themselves prefer Western goods to Asian ones. There is a general Asian impression that the goods designed and labelled as coming from the US, Japan and Europe are better than Asian ones.

Let us do a small thought experiment. For a moment, imagine yourself staying in a five-star hotel in the centre of a sophisticated European city – Rome. The hotel costs about €300 a night. You go into the room, which looks attractive (just as you had expected it to be in Italy), with its vibrant colours and ornate furniture, besides being well equipped. Slowly, you start looking at the articles in the room and you realize that the clock is made in Malaysia, the furniture in the Philippines, the hairdryer and the TV are made in China and the curtains in India. How do you feel about the hotel and its quality? Do you feel happy about paying €300 a night? Do you feel a bit cheated, as if it were not a five-star hotel? Is it because the hotel is using *low-cost* Asian supplies? Is this what you expected in a top-quality hotel in Rome? Next time you walk into the lobby, do the bright colours that were pleasant earlier begin to appear a bit gaudy and cheap and does the furniture start looking odd and a bit unfashionable?

Why do made-in-Asia articles not bring a sense of pride to Asians? Asians frequently sell themselves short in their own minds. In general, if an Asian can afford a Western brand, how often does he or she prefer to buy an Asian or domestic brand? Even in relatively 'nationalistic' South Korea, the Western garments are often preferred over those by local companies. It isn't that Korean consumers cannot find anything Korean that is good or suitable, but to find respect in society, they often prefer to buy expensive Western brands. For instance, in a study of the passiveness of Korean consumers and their obsession with Western goods, a young student from a top private university was quoted as saying:

> When I saw a bag with the brand name 'Jansport,' I did not buy it, because I thought it was a cheap Korean imitation of an Eastpak [a back pack]. But when I discovered that Jansport is also an American brand name, I changed my mind and now I find the product very attractive.[5]

It is surprising that, while Americans often talk of their US brands, Europeans talk about theirs and the Japanese prefer their own brands, consumers from the rest of Asia prefer American, Japanese and European products!

Think about the advertisements you see in Asia, especially Southeast Asia. Almost every other advertisement has white faces in it. Malaysia is an exception and was criticized heavily in the international media when the use of a poster featuring Brad Pitt was forbidden by the government and the then Prime Minister Mahathir suggested that Asians should try to use Asian models in their advertisements. Domestic Asian companies selling goods to domestic consumers with Caucasian models – seems strange but, that is what happens. Many people are surprised when they find out that Esprit, which uses primarily non-Asian models in its advertisements, is a Hong-Kong owned brand. There seems to be so much talk of localization but it is not evident in Asian advertisements!

One Filipino air-conditioning company even has a Japanese name to create a perception of a higher degree of sophistication. Lenovo and many other Chinese companies want to have a more 'neutral' name lest their Chinese image has a negative impact on international sales. Ever heard of a European or American company trying to disguise where they are from? Is this flexibility or a lack of self-esteem?

Why do Asians behave this way? Kishore Mahbubani of Singapore calls this Asian phenomenon 'mental colonialism'. Just as the traces of colonial mentality are difficult to pinpoint, so is mental colonialism. It can be observed but is difficult to scientifically verify. As a simple test, you could make a careful observation of people around you in Asia. We will be surprised if the results differ, although it is possible.

One other plausible explanation is that a significant percentage of Asians tend to co-relate self- and others' 'worth' with financial assets, and therefore tend to look down on their own companies and their fellow Asians when comparing them to companies and individuals from the industrialized countries. Since it easier to follow a crowd than create and lead one, Asians tend to follow the 'fashionable styles' from Japan, North America and Europe rather than trying to assert their own indigenous values and preferences. They also feel that it is difficult for them to create international 'fashionable styles'. For the younger generation, the 'coolness' factor of the US and Japan also holds some explanatory power.

What happens when such consumer behaviour exists in an economy? The lack of customer support domestically or in the region makes it difficult for Asian companies to grow themselves in domestic or other Asian markets while projecting themselves as Asian companies. Obviously, this has a significant feedback effect for the aspiring innovator, considerably worsening the situation.

Can you do something about it, without having to wait for a total change in the mentality of the Asian consumer or buyer? Once again we found an example. A relatively small company in India has a software solution called Tally. This is the colloquial term used by the local accountants when they talk about cross-checking accounts. Tally has been popular and even those accountants who avoid using computers have found it very convenient. According to research done by the International Data Corporation (India), in 2004, Tally was bigger than all the other accounting software brands in India put together. The reason was quite simple: Tally's flexible and easily comprehensible business solution mimics the human train of thought and using it is as intuitive for accountants as the manual accounting system. Originally made for small businesses, today Tally is used by businesses of all types and sizes with over 1 million users in over 88 countries worldwide. In India, which is known for its software services and *not* software products, Tally has been hugely successful. And the credit goes to the attention paid to the style and tendencies of local accountants and then offering the software solution at an attractive price point, neither of which the big brand software companies from the West could do. This example is more than simply one of creating a better image for a local product. It is also about addressing unserved markets (as discussed in

Chapter 5). But like other, more well-known examples such as Singapore Airlines in the airline industry, or some of the retail brands in the clothing industry, it does show that a local company can make its products the preferred choice for Asian customers.

SUCCESS IN MANY INDUSTRIES REQUIRES ACTIVISM IN SETTING INTERNATIONAL STANDARDS

In a world of shortening product lifecycles and rising costs in bringing innovations to market, rapid development is important. When it comes to developing products and services that require the participation of other vendors/ service providers, it is imperative to agree on standards so as to create a win–win position for all interested parties. The standards we refer to here are the uniformity/technical standards that allow interoperability of various parts. This reduces reliance on certain suppliers and gives freedom to customers and manufacturers alike to interchange components freely without being tied to a particular vendor.

Generally, global industry standards are not *legally binding* standards as such. A variety of companies form consortia to develop standards. Sometimes, there are competing standards formed by different consortia. The primary constituents of consortia are commercial organizations with vested commercial interests. To protect those interests, a significant degree of lobbying takes place in these consortia. While what finally emerges is called a 'standard', it is often a negotiated outcome after significant 'horse trading' has taken place among the key players who are often none other than the traditional dominant players in Europe, the US and Japan. For example, the telecom standards are set at ITU in Geneva. The standards for an operating system for a PC are, for all practical purposes, determined by Microsoft in a quasi-partnership with Intel. The standards for CD players are determined to a large extent by the Japanese consumer electronics companies. When standards are set by consortia, the meetings generally take place far away from Asia in the major European and US cities (sometimes in Japan) and are chaired by companies' representatives from the major players. Often, the headquarters of the global consortia are also located in the US, Europe or Japan, affecting the ability of Asian companies to lobby effectively. With their considerably smaller resources than those of the established global players and their 'follower status', Asian companies generally have little or no ability to influence the agenda and opinions of consortia even when they are invited to participate. Many Asian firms themselves do not take an initiative either because of the self-perception that they 'do not count, really!'

It is surprising that even in industries like telecommunications and IT where Asia is well developed, with a large number of suppliers and a fast growing market with significant potential, until recently Asian companies have had a limited sociopolitical influence. And, when it comes to taking leadership centre stage, Asian firms have had an even more dismal record of success. One is hard-pressed to find any technologies of significance where Asian firms have taken the central leadership. Even Japan's home-grown cellular standards and technologies (PDC and i-mode) have failed to win general acceptance outside the Japanese market.

The fact that Asian companies do not play any role in developing and setting industry standards is hugely detrimental to their future and their ability to extract profits out of their innovations. Without influence over de facto industry standards, Asian companies cannot effectively bring to market their own innovations which require complementary products and services. Economies of scale deter the suppliers from supporting the companies producing non-standard goods/services.

As the future depends on its sources in the past, playing second fiddle will only force Asian companies to continue to be dependent on others' standards. Breaking the vicious circle of non-significance in standard making is important for Asian firms. Awareness is the first step, and action is the second.

In November 2003, China started to exert influence for its firms. It announced the prohibition of the import or sale of certain types of wireless devices that did not comply with a domestic security standard known as WAPI. The incident sparked outrage among foreign-invested equipment makers and a sharp response from the US. In a recent letter to the Chinese government, the US government accused China of creating a dangerous precedent for using technology standards as a barrier to trade. An English magazine *Business China* reported in March 2004:

> On the face of it, there is nothing wrong with Chinese consortia co-operating to develop new standards. The effort may be wasted if the new products cannot compete, but that is what competition is about. Nor is there anything unusual in governments cheerleading for domestic companies, as US support of Qualcomm's CDMA technology demonstrates. But Beijing crosses the line when it tries to impose standards by requiring consumers to use them or manufacturers to produce them. This is especially true in a vast market such as China where whatever becomes the norm quickly generates economies of scale that could be transferred to markets across the world.

Economies of scale lead to one of the most important benefits of the power of standards. Asian companies need to do something to modify their past non-participation in standard making.

China has been trying to do that with other technologies as well. In the case of 3G rollout in China, there appears to be a delay, because China is trying to back a local 3G standard known as TD-SCDMA. Given the likelihood of its success, the GSM Association, representing about 70 per cent of the world's 1.3 billion mobile phone users, has already made a cooperation agreement with the Chinese standard to coordinate the development. China has domestically embraced a unique DVD standard (known as EVD, or enhanced versatile disk) in an attempt to avoid royalty payments that would otherwise be due to holders of DVD patents in Japan, the US and Europe. Several additional programmes aim to develop products that will bypass the need for royalty payments to foreign companies, including new standards for optical storage and electronic displays, digital TVs and a TCP/IP protocol known as 3C that will allow consumer electronics, computer and communications devices to talk to each other. Indeed, there is significant resistance in giving China an opportunity to make these standards international.

Without the collective efforts and/or government support, only a handful of Asian firms have the ability to influence standards internationally. Think about operating systems or desktop application software. Towards the end of 2003, the governments of Japan, China and South Korea and the leading IT companies in those three countries got together in an initiative to develop an open source operating system (building on existing Linux) to have lower cost, reliability and full control. From their perspective, Microsoft Windows has dominated everything, even though people also want to test different products. Therefore it is important to work on alternatives. This partnership is expected to develop open source business models, standardize software and train software engineers. The partnership has already moved towards a division of labour: China will develop PC operating systems, Japan will focus on software development and security, and South Korea will develop software for PDAs. The partners are setting up a database to coordinate their efforts and avoid duplication. Such initiatives may fizzle out after a while, but this is a serious attempt and the initiative is to be guided by coordinating bodies with considerable clout: the Japanese IT Services Industry Association (JISA), the Chinese Software Industry Association (CSIA) and the Federation of Korean Information Industries (KFII). The three associations have more than 1000 corporations on their membership rosters, including most of the technological heavyweights in the Japanese and Korean industry. The Japanese newspaper *Asahi Shimbun* wrote that 'the plan is aimed at "smart devices" such as mobile telephones, digital cameras, car navigation systems and computer servers. The idea is being driven by a fear that Microsoft could corner the market for such devices as it has done on the desktop'. The intention is clear: not to give up the Asian IT innovation to Microsoft's standards.

Asian firms have an interest to support and actively lobby their governments to be involved in standard setting. But for a small firm, that is not always a feasible strategy. There is another one: if you can't beat the standard setters, join them. At the firm level, iRiver and some other Asian firms partnered with Microsoft in an initiative to build a new hand-held, video jukebox called the 'personal media centre'. With this alliance, they are trying to create standards that can be pushed in the market. iRiver, for instance, is not only developing content that can be used by all devices with a Microsoft platform, but is also involved with Microsoft in fine-tuning the digital rights management techniques.

Standard making gives a lot of power. Asian firms need to actively participate in standard making to start securing their future. The participation efforts should be treated as investments in their future. Over time, standards become so sticky that is almost impossible to start something totally new.

PERCEPTIONS ARE CREATED BY THE MEDIA: MANAGE THEM!

We are all influenced by the media. Since it is impossible for one to experience and evaluate everything first hand, one has to rely on the media as secondary sources of information. Various forms of media such as business newspapers and magazines are perhaps the most influential in shaping the perceptions of consumers and the people involved in business. It is thus unfortunate that international business media reports are sometimes negative or create negative perceptions about Asian firms.

Media sells news; news is interesting when it is sensational. Often, the business journalists are generalists and have little specialized knowledge of business functions. But a significant number of readers tend to interpret their views and articles as if the reporters were experts. Many journalists, partly due to the limited resources at their disposal, seldom intellectually engage in verifying the facts obtained from the PR officers of firms and the 'subject matter experts'. The reporters and editors of the Anglo-Saxon business media often have an ideological view of the moral and practical superiority of Anglo-Saxon-style capitalism. They highlight and wish to explore only the issues (for instance accountability and, lately, corruption) with the contemporary lens of the Anglo-Saxon world, without regard for the social, political and economic constraints and the level of development of the countries they are reporting on.

What the media chooses to report is important because it tends to focus readers' attention and create impressions. More than six years after the Asian financial crisis, a bad image of Asian companies still lingers on. Except for the occasional article, until 2003 the English business media continued to be somewhat negative about East Asian firms. This is in sharp contrast to the dominantly positive tone before mid-1997 when the media fanned the emerging belief that southeast Asian firms would take over the world and suggested that they might even have superior models of management. In the past, the media had said similar things about Japanese and South Korean firms. As such, due to the nature of their own business and revenue generation, the media tend to focus on reporting what affects their key advertisers and their customers. [6]

The business media can play an important role in the success (or failure) of economies and their firms. In 2003 the India Brand Equity Foundation reviewed a select set of publications,[7] based on their popularity and ability to generate public opinion in key global markets, including, but not limited to, the USA and the EU. The total number of articles published about India (and not just on business and economy) in the tracked magazines was *only* 213 (as you might imagine, China's score is many times higher). Of these:

- 88 were positive
- 56 were negative
- 36 were general articles
- 21 were general articles with negative undertones
- 12 only were general articles with positive undertones.

Such treatment has created different interest and had a different impact on the Chinese and Indian economies. This can partly explain the differences in international public awareness (and action) about China and India.

In terms of tone and the choice of words the media, often unknowingly, projects a condescending view of the emerging countries in general. Think about the word 'outsourcing' – it often has the negative connotation that low-value, low-price work is being shifted to Asia. However, the literal meaning of 'outsourcing' is similar to the concept of focusing on 'core competence'. Asian parents do not talk about 'outsourcing' the higher education of their children to US universities – do they? And think about the term 'telemarketer'. It is commonly understood and used even today. But, for telemarketers in Asia, the term used is 'call centre workers'.

All this creates a negative image of Asian firms and the impression that Asian firms create low value even if they do the same or similar work as that done by many firms from the developed world. The English-educated Asians in Asia who are also regular readers of the English international business media also begin to develop poor impressions of firms in Asia and tend to shy away from participating in them. But, Asian firms and Asian economies need the contribution of these people for their growth.

Readership and influence of the local Asian press are limited in international circles. Although the situation is changing, censorship (self- or enforced) has been widely practised in Asia until recently. The quality of reporting has often left a lot to be desired. Low pay for journalists and a focus on sensationalism rather than on facts has also been a feature of the media in many Asian countries, at least until recently.[8] Thus, the Asian media lacks the credibility required to make an impact. Of course, there are exceptions. Certain local journals are considered to be of high quality, but they are rare.

In the meantime, the firms have to find a way to avoid the negative stereotyping that has been created and is reinforced regularly. There are some companies that have found creative approaches that have worked well for them. For instance, Bank Mandiri, currently the largest bank in Indonesia, was formed during the Asian financial crisis in an attempt to restructure the crisis-struck banking sector. Mandiri is the result of a merger of four large bankrupt state banks of Indonesia. As can be imagined, everybody in the country had a poor opinion of state banks at that time. The IMF did not trust the government, the government did not trust the Mandiri management and the Mandiri management was forced to accept the 'advice' of the foreign advisors engaged by various multinational institutions. Within three years of the official merger in 1998, considerable progress had been made but the bank was still perceived poorly within and outside Indonesia, which crippled its growth plans. It invited researchers from INSEAD, a leading international business school, to write elaborate case studies on the merger and transformation of Mandiri. The academic case studies show an unbiased true picture which perhaps the media would find difficult to show. The INSEAD case studies (which also contain a fair bit of criticism of the management) have been used by Bank Mandiri to showcase itself to investors and other stakeholders. Thus, case studies by leading business schools have been a good source of painting an unbiased picture that is trusted in the market. Other Asian companies such as Acer have used this method to gain the attention of the international audience.

Another method that has been used by companies is to get their CEOs to be speakers or panel members at conferences, giving the CEOs the opportunity to talk directly about their own company and create awareness. More importantly, if the CEO or other senior executives can start to take on the role of spokesperson for a rising industry or talk about something that is of general importance to the business community, he/she attracts attention from journalists who would look into the company and have a more positive frame of mind. Kiran Mazumadar of Biocon in India has done just that. Although until a few years ago, she ran a relatively small biotech company, she took every opportunity to be on stage and be a spokesperson for the industry, eclipsing the CEOs of larger corporations. The journalists started taking note of her company and the industry at a time when the focus on India was its IT industry. Similarly, Napoleon L. Nazareno of Smart Communications has put his Filipino company on the global technology map. Having built innovative mobile systems and services in the Philippines, without 3G, he has gone on the international speaking circuit to surprise and educate the world about the innovations in the field in the Philippines. He is a regular speaker at top industry conferences .

Thus, firms in Asia need to put in extra effort to manage their media and public relations to help capture better value for themselves.

CONCLUSION

In our research we discovered, somewhat to our surprise, that a group of Asian companies find it difficult to innovate because of a perception, a feeling, that they did not have enough capabilities to innovate. This is the result of many subjective and objective factors, among them the negative perception of Asian products and services in the eyes of both Asian and other customers, the absence of influence in standard setting and the critical, if not negative, image about Asian firms that often is projected by the international media.

Many of these factors will not disappear easily, and require collective efforts by firms and governments. Images do not change quickly and only a long-term effort will lead to a real change. But firms cannot wait until that change has happened and they need to take action now to break out of the vicious cycle of the underdog mentality. Recognizing this is a first step. Doing something about it is the second.

Call it marketing, call it public relations, but what we see in many of the examples in this chapter is that senior managers and CEOs have embarked on an educational exercise. Whether it is Yeap Swee Chuan from Aapico who invites senior managers from European firms to come and see his operations for themselves, Bank Mandiri that builds up a relationship with internationally recognized researchers or Kiran Mazumadar of Biocon who acts as a spokesperson for an industry, it is obvious that top management sees it as its responsibility to be personally involved in overcoming the negative perception that may exist about the company's Asian image. This is perhaps the most important lesson of all the examples: it is top management's responsibility to break the negative impact of the underdog mentality.

Notes

1 Global Directions, *The McKinsey Quarterly*, 2003, (4).
2 Personal interview.
3 Ruimin Z., 2000, Innovation Makes Haier Strong, *China Today* **49**(9).
4 The Capability Maturity Model (CMM) from Software Engineering Institute (SEI) of Carnegie Mellon University has been a model used by many organizations to identify best practices useful in helping them increase the maturity of their processes. Level 5 is the highest level of maturity.
5 Duk-young K. Subjectivity, meaning and culture (Naman, 2001) – quoted in Passive Korean Consumers Follow Behind, *Korean Herald,* 26 June 2004.
6 For detailed treatment of this matter, the reader may refer to Chomsky N., 1989, *Necessary Illusions, Thought Control in Democratic Societies*, South End Press, Boston.
7 *Time Magazine* (Asia Edition), *Fortune* (International Edition), *Business Week* (Asia Edition), *The Economist* (International Edition), *Newsweek* (International Edition), *The Far Eastern Economic Review*.
8 'Shortcomings of the media in Asia', Political and Economic Risk Consultancy, June 1998, http://www.asiarisk.com/library9.html.

9 What to do next?

We have come to the end of our endeavour. It is time to review the objective we set ourselves at the beginning of this book. From any recent analysis of the competitive landscape in Asia you will have learned that some major changes are taking place. The emergence of China as a leader in manufacturing, and in the near future in design and development, and the rapid rise of India as a provider of IT and IT-enabled services, change the macroeconomic constraints under which other Asian firms have to operate. Competing on low-cost production and low sales prices is becoming increasingly difficult for firms from other parts of Asia. Also, within India and China the low-cost/low-price competition is constantly renewed and is moving inwards and towards more rural areas. For many firms the only effective response to this change in the competitive scene is to move up the value chain. That is what firms from the industrialized sectors such as Japan, Western Europe or the USA have done in the past. This is what Asian firms need to do today. Moving up the value chain will require more innovation. Innovation in products, services and processes is the only effective answer to this new competitive situation.

How do you innovate? Is the approach you use in Asia different from the one you would use if you managed a company in the Western, industrialized world? Are there any special tricks that one should use in the Asian environment? Those are some of the questions we set ourselves to answer through a combination of case studies and a survey.

In this chapter we give a short overview of the results of our research, interspersed with questions that may guide you in the development of a better way of managing innovation in your organization. Take the time to answer these questions. Taking this time is important. Answering these questions will give you the real return on the investment you made when you decided to read this book. Some of you can answer yourself, while others may need discussions with your colleagues and thus may take a bit longer. Sit together with your colleagues and discuss innovation in your organization. Organize a workshop. We are convinced that the exercise of answering the questions will help you to focus on and improve your approach to innovation.

DISCOVERING THE HURDLES TO INNOVATE

Early on in our research we discovered what many would have predicted: the basic principles of innovation management that have been developed over the past 30 years in the US, Japan, the UK and Continental Europe do apply in Asia. We summarized these principles in eight categories in Chapter 2. Just to remind you of them, we have reproduced them here:

1. There is no innovation without leadership
2. Innovation requires calculated risk management
3. Innovation is triggered by creativity
4. Innovation requires organizational integration
5. Success in innovation requires excellence in project management
6. Information is the crucial resource for effective innovation
7. The results of creative efforts need to be protected
8. Successful innovation is rooted in a good understanding of the market.

Question 1: Do you apply all eight principles in a logical way in your organization? Which do you think your organization applies? Which are you not good at?

But, as with the implementation of any basic principles, there is perhaps some localization needed in adapting them to the typical environment for Asian firms. And there is some further localization needed on how you apply them in your own specific situation in Asia.

This book has come up with five broad areas that need localized approaches. Through a wide range of case studies in small and large product and service organizations and high- and low-tech organizations, the following five hurdles emerged:

1. Technical expertise and risk capital are both needed for effective innovation but are still scarce throughout Asia.
2. The markets needed to stimulate innovation are either far away from Asia (either geographically or culturally) or, if they exist locally, are often too small.
3. The existing industrial policies of most Asian governments are too often about catching up with industrialization, and do not sufficiently aim at value creation through innovation.
4. Many organizations in Asia have a risk-averse organizational culture. As a result, they limit creativity and create an underdog mentality of not being able to innovate.
5. In Asia there is a considerable lack of appreciation for intangibles, for example intellectual property rights or brands.

Question 2: Do you recognize some of these categories as being a problem for your organization? If so, have you been able to overcome any of them and how did you do it? Which one do you consider to be the most important? Be specific in your answers.

NOT ALL MANAGERS FEEL THE SAME ABOUT THESE HURDLES

Through a web-based survey we confirmed that these five areas were indeed relevant to Asian managers. Of course, they may to some extent be relevant in other areas of the world but we do not have empirical evidence to support that. We also found that not everybody in the sample had a similar sensitivity to each of these areas of attention for localization. We will not go into detail here but you can go back to Chapter 3 and reread the list of 32 items in Box 3.2 and the ten dimensions along which the managers who responded to the questionnaire differed in their evaluation of the importance of the many hurdles that may exist to hinder innovation.

Here we will cover the five most important ones, that is, those that explain the 40 per cent variance in the answers. The most important factor reflects the *absence of an environment in Asia in which it is easy to operate as an innovator*. This combines the lack of good market data and the lack of trendsetting customers with the stigma associated with failure. The second factor groups those items that reflect the *underdog mentality* of the Asian company: to what extent are Asian companies involved in setting international standards, do they lack confidence, do they have good market perception and do they see the international business press as constantly negative about Asia? The third factor is the *lack of some knowledge resources*, for example the lack of high-quality knowledge workers and the lack of reliable competitive intelligence. The fourth factor points at the *inertia created by the forces of tradition in Asian business and management*. The fifth factor groups those items to do with the *lack of basic management models and management lessons specifically applicable to innovation management in Asia*. This is the factor that cries out for the localization of management methods to manage innovation in Asia.

The managers who responded to the questionnaire 'scored' differently on these factors. But we could group the respondents together in roughly equal sized groups. In fact we discovered four distinct groups of managers. Each group seemed to be preoccupied with a different set of challenges and therefore they may be interested in different types of solutions for their organizations. The four groups are:

1. The *innovation starters* are those managers who do not suffer from an underdog mentality or the inertia resulting from a traditional Asian management style. But they don't feel the need to innovate yet and also

complain about not having the appropriate management methods. We chose this label because we suspect that this attitude is typical for managers of companies who have not yet committed themselves and their organizations to innovation. If you recognize yourself in this group, you may want to reread Chapter 2 to get a feel for the basic management lessons about innovation. You will also benefit enormously from many of the examples in this book because they will provide you with ideas on how to start the effort to innovate.

2. The second group is called the *tradition fighters* reflecting those managers who recognize that the lack of knowledge resources, as described in Chapter 6, or the lack of a perceived need to innovate is *not* their problem. They know that they can and need to innovate and think they have the resources to do it. But in their momentum they seemed to be stopped by the inertia created by traditional Asian management methods and organizations and the underdog mentality described in Chapter 8. We labelled them tradition fighters because of what they see as their challenge: to fight traditional approaches and renew their organizations. If you recognize yourself in this group, you may benefit from going back to Chapters 4 and 8, which give a range of stimulating examples on how to change the traditional organization and overcome the underdog mentality.

3. The third group is those managers who perceive they need more knowledge resources, so we called them the *poor in knowledge resources*. They seem convinced that they know how to manage innovation, and probably most of what we described in Chapter 2 on the basic lessons of innovation management was well known to them. But their challenge is to mobilize the resources. The examples in Chapter 6 will stimulate you if you identify with this group. But also the ideas on profit management in Chapter 7 will be useful because they will inspire you to make the most of the limited resources you can deploy.

4. The fourth group brings together those managers who have less of a problem with the availability of resources and appropriate management methods and do not feel that they are paying too much tribute to a typical cost reduction attitude that would hinder innovative endeavours. So they really think they can and should innovate. But they suffer from an underdog mentality, do not see the rewards for innovation and blame the government for not creating the right environment. We called this group *stuck in the muck*. This label is a bit tongue in cheek: some of these managers may blame the external environment too much for their lack of success with innovation. Our advice here would be to get on with the task. You have the resources and the capabilities. Go back to Chapter 8 on how to overcome the underdog mentality and ensure that you apply the suggestions from Chapter 5 on markets and marketing. The markets can be a trigger to mobilize the energy to get out of the muck.

Question 3: Do you identify with one of these groups, or with a combination of them? If not, you may already be a leader in innovation and you can take the examples in this book as a benchmark for continuous improvement.

Whichever of the four types you feel close to, we are convinced that some of the examples we offered in Chapters 4–8 will stimulate your thinking. As we mentioned at the beginning, it was not our intention to provide you with dogmas or prescriptions on how to manage innovation. That would have been presumptuous on our part. Firstly, most of the general lessons do exist and can be found in the innovation management literature. And secondly, it is precisely in the implementation of these general lessons, that is, the topic of this book, that creativity can make a difference between failure, success or enormous success. Rather, we hoped to inspire you and trigger your own creative thinking. Looking at examples and access to information and ideas will render you more creative.

Stimulating your creativity was what we hoped to do, so we used a wide variety of examples. It would have been simple for us to limit ourselves to the icons of success in Asia, for example Samsung, Singapore Airlines, Creative Technologies and so on. But you know many of these examples anyway. What we wanted to show was how you can find creative suggestions in some of the lesser known companies, even in those that may not always be successful in the long run. There is an important message we wanted to share with you in choosing examples from these companies: good practice and inspiring ideas may come from unexpected corners and the successful innovator is usually on the lookout for these lesser known gems.

Question 4: Are you alert to ideas and best practice that may be learned from companies other than the traditional icons of successful management in Asia?

A QUICK SUMMARY OF OUR OBSERVATIONS

A short summary of this book is unfair to the companies that we have described in some detail throughout the book. You really should go back to each of them in view of what you found most inspiring. But we will attempt to provide you with a quick summary. Rather than summarizing each chapter, we decided to bring together some inspirational ideas from the previous chapters under the eight management lessons of Chapter 2. This will make the final link between what is applicable worldwide and what is needed for broad localization in Asia. You will notice that some principles are discussed at greater length than others. This is because in some areas, Asian firms have already built up a lot of experience. You will also notice that we pay most attention to the issue of leadership, one of the most important challenges for senior managers who want their organizations to become innovative.

There is no innovation without leadership

The leadership that is needed in Asian firms is often about how to instil a belief that innovation is possible. The most important lessons on this were highlighted in Chapter 4. In this chapter we showed that you have to create an attitude that is different from the traditional management wisdom in Asia. There is some 'unlearning' needed. Future competitiveness will not be built on cost competitiveness alone. Only higher value will command sufficiently high prices to create a sustainable competitive advantage. Success will need to find some of its roots in the creation of knowledge, that is, actionable information that we believe to be true is the core resource of the organization.

Some Asian organizations and firms have already made this mental step and we can be inspired by some of the actions they took to implement the new organization. Their organizations have invested in the creation of an environment that stimulates process improvement and knowledge creation. Aapico Hitech is an example of how this change in attitude was achieved over a period of several years.

The change in attitude needs to happen at three levels. Firstly, Asian company leaders need to ensure that their organizations embrace value creation. This value creation will require a strong entrepreneurial spirit and a willingness to take calculated risks. But that will not be enough. At the operational level, managers must move from creative improvisation to solve problems, towards careful and well-thought-through process management in both design and development. At the strategic level, Asian company leaders will need to share a vision about their firm as a bundle of capabilities, that can be constantly recombined in order to respond to market needs. Companies are not portfolios of products. It is the portfolio of capabilities that makes a firm.

Some Asian companies feel constrained in their efforts to become more innovative because of their own perception that they don't have what it takes to be a leader in innovation. They seem to think that they lack the necessary capabilities. There are some good reasons for this perception. Many firms, as we saw in Chapter 8, struggle with a negative image of themselves and their products in the perception of both international and Asian customers. They are also at a disadvantage because they do not participate sufficiently in international standard setting. They need to become more proactive in these international forums. And they need to work on the international process, in the hope that this will lead to a more positive reporting about Asian business and firms.

Changing the mindset, changing the vision and strategy, and becoming more assertive on an international level will take quite some time. An image does not change overnight and only a commitment to change over the long term can lead to real change. The commitment will have to be a collective one. An isolated firm will have difficulties changing the view the world has of an Asian company. Collective efforts may pay off faster. That does not mean one should be passive and wait until governments or industry organizations take action. Firms need to initiate change. They need to take action to break out of the vicious cycle created by an underdog mentality. In Chapter 8 we gave the examples of Aapico, Biocon, Bank Mandiri, where their CEOs took it into their own hands to ensure that their organizations improved their self-confidence. This type of leadership is needed in Asian firms, perhaps more than in Western organizations. You will not find this need for increasing self-esteem in a traditional text on innovation management. But for innovation in Asia it is a requirement.

Question 5: What did you put in place to provide this leadership? What mechanisms described in Chapters 4 and 8 can you apply in your organization?

Innovation requires calculated risk management

In Chapter 6 we argued that Asia lacks sophisticated risk capital and capital markets that can evaluate risks in a calculated way. This creates an additional cost for the innovator. But we should not be naive. The situation is not going to change overnight. Therefore the Asian innovator will need to be creative in finding additional sources of financing. As we illustrated with the examples of Patkol, Aapico and Netizen Funds, there are creative ways to either reduce the need for risk-taking financial resources or find these resources in places outside your own country.

Question 6: How do you reduce the need for capital in your organization?

Innovation is triggered by creativity

Creating an environment that enhances creative behaviour also rests on the use of three levers. The first one is the creation of more cultural and cross-functional diversity within the management team. Asian companies may still have a lot of difficulties with that. They often prefer homogeneous environments. But well-managed diversity stimulates creativity and enhances performance. The second lever is to increase the motivation of your employees through more professional management, appropriate incentives and in particular the decentralization of initiatives throughout the organization. The third lever is to apply an appropriate dose of stretch to render people more creative. But don't forget what we saw in Chapter 2: too much stretch can also be paralyzing. Capital markets and in particular careful exposure to best practice can help you to realize this stretch.

In several chapters we also touched on the subject of learning. When we discussed the challenge of markets in Asia in Chapter 5, we pointed out that large markets with low spending power can sometimes be a source of creative learning. And in Chapter 7 on profit management we advised you to see imitators as a source of learning. Both these ideas can help to stimulate your creativity in innovation.

Question 7: Would it not be worthwhile organizing a discussion or workshop with your colleagues to come up with some alternative ideas and write an action plan as to how to achieve them?

Innovation requires organizational integration

The challenge of organizational integration is not exclusive to Asian firms – perhaps large non-Asian firms suffer even more from a lack of organizational integration. There are two observations we can learn from the examples in this book.

First, the traditional structure of the Asian family firm leads to a fairly hierarchical, top-down form of communication. Under some circumstances this is an effective way of organization but it does not stimulate creativity, and the top can become a bottleneck in decision making. Asian firms will have to learn more about flatter and networked organizations in order to embrace innovation.

Secondly, we noticed that with the growing internationalization of Asian firms, there is a growing diversity in the workforce (not in the management yet). This helps in stimulating creativity and gives access to different market signals. But it can lead to organizational disintegration if the interaction between diverse groups of employees is not managed carefully. Growing internationalization can lead to more innovation, but does not guarantee it. Senior company leaders will have to pay a lot of attention to the global integration.

Question 8: How can you move towards a flatter and more networked organizational structure in that part of the firm where innovation is needed most?

Success in innovation requires excellence in project management

This is probably the area where this book contributes the least ideas and there is a simple reason for this. Operational excellence is what many Asian firms have invested in over the past decade, and they have mastered the challenges of project management. Just because we did not touch upon this very often in this book does not mean that it is not important. We are firmly convinced that excellence in project management is one of the eight key elements in successful innovation management and constant follow-up and improvement in project management will pay off handsomely.

We did, however, stress and illustrate in Chapter 7 that speed in execution, that is, the result of excellence in project management, is a key element in profit management.

Question 9: What would it take to shorten the development time of your next product or service by 33 per cent compared to the previous development?

Information is the crucial resource for effective innovation

Innovation requires knowledge resources. Information consists of two types: codified and non-codified. Asian firms can easily gain access to codified knowledge. But to innovate you often require a lot of non-codified knowledge. This non-codified knowledge mainly comes through people. The scarcity of good technical and human resources in Asia is perhaps an important hurdle for this transfer. Without the appropriate human resources, creative entrepreneurs will not be able to achieve their ambition. In Chapter 6 we drew your attention to the fact that the momentum of Asian innovators may well be stopped if they do not have the quantity and quality of technical infrastructure, engineering, design and management skills, and creative employees.

The availability of these resources depends on government initiative. We expressed the hope that government leaders will put in place the systems to develop these resources. But don't forget that managers can also take some action to overcome the temporary shortages with which they will be confronted.

As we described, engineering and design skills can be developed inside the organization. Managerial skills can be kept at a high level by maintaining good networks with your friends and peers in industrialized countries. Other examples showed how alliances or acquisitions in industrialized economies can also be a good policy to gain access to these skills. But as we pointed out, these alliances or acquisitions may be difficult to manage.

Question 10: Can you come up with one action to increase the pool of well-educated designers in your organization?

The results of creative efforts need to be protected

One of the basic premises of this book is that beyond the act of invention and launching a product or service you need to invest in profit management so that you get the rents out of the effort you make as an innovator. You may remember from Chapter 2 that we consider an innovation only to be successful if we have an economic success, that is, a return that includes a reward for the risk that innovation entails.

Getting to such returns requires careful profit management. In Asia entrepreneurs' profits often disappear because of copying. In Chapter 7 we highlighted the fact that illegal imitation is not only rampant in Asia, but is harmful for Asian consumers, governments and innovative companies. A first action plan is for Asian governments to invest more in intellectual property protection. Indeed, this will take time to implement. And there may be some good reasons why some governments will drag their feet. Therefore companies have to engage in strategies to cope with imitation.

Pursuing the enforcement of one's legal rights can be one way. But experience shows that this can be costly and it may take forever. Therefore it is not clear whether the owner of the IPR can really achieve a worthwhile return out of it. Perhaps the approach should be to try to obtain enforcement of legal rights, and combine this with a more managerial approach. The two major suggestions for inspiration that we gave in this book were: learn from what your imitators do and manage as if all imitation is legal. Consider that IPR is a windfall profit and thus you have to defend yourself with other competitive means. Indeed, there are examples of companies in Asia that have pursued these two approaches. The speed with which Smart Communications from the Philippines innovates and the co-creation efforts by iRiver that engender a high level of loyalty with its young customers are two interesting examples of how companies have outpaced their competitors, and are described in detail in Chapter 7.

Question 11: When you consider all imitation to be legal, what kinds of actions can you take to increase the time that you keep a lead over your competitor, without reverting to patents?

Successful innovation is rooted in a good understanding of the market

Innovation without intimate customer knowledge is going to be difficult. We do not deny that in some rare cases a breakthrough innovation is the result of insight by a brilliant and visionary innovator who can imagine a need before the customer or the market can do so. But in this book we focused mainly on business innovations in products, markets and services. Those innovations require an intimate understanding of the customer and the user.

We argued that Asian markets are not yet the most supportive for an innovator. Customers tend to be more conservative, markets are heterogeneous and market data is not available, and the most sophisticated markets are far away, geographically and culturally. But the examples we gave of companies like AmorePacific from South Korea and Haier from China and many others suggest that there are creative ways to overcome the hurdles and the challenges.

Sophisticated markets will always remain an important source of learning. Therefore innovators will need to set up antennae to tap into these sources. There are several possible approaches, ranging from the creation of a wholly owned subsidiary, setting up an alliance or acquiring an overseas unit. None of these strategies is easy to implement, but the examples of Haier from China and Samtel from India show that it is possible for those companies that have the tenacity and patience to succeed.

Asian markets are often considered to be poor and thus uninteresting for innovators who want to launch new products and services at premium prices. But these low-income markets also tend to be large. Smart Telecommunications and Tiger Motors are two examples of companies that have used their creativity to customize innovation for mass markets instead of just focusing on the crowded premium segments. The size of the low-income market can make it very attractive, even though one cannot charge premium prices and make good margins.

It may be tough to find early adopters, for cultural or other reasons. But as the examples of Hindustan Lever, iRiver and Patkol show, it may well be possible to use indirect ways to seduce lead customers to adopt the innovation and become your best ambassador. Backing this up by a clever communication policy, often involving the early adopters, and guaranteeing the availability of the product or service can help.

Finding marketing data is tough. So, experiment. Heterogeneity needs to be overcome. Awareness of this paucity of data and the differences between Asian countries is the first step. After that you can find ways to engage customers, share the risks created by the lack of information and set up careful experiments to learn.

Together with your colleagues, develop a concrete action plan to develop one lead user for your next product. Also start an exercise to figure out how the vast market of consumers with low spending power could be a market for you. This does not mean that you necessarily need to go into that market, but the exercise may reveal lots of good ideas for your traditional markets.

AND NOW IT IS UP TO YOU

In writing this book we wanted to inspire you. But, without action, inspiration leads only to intellectual satisfaction. We hope that some of the ideas in this book will lead to action by you. We hope that by showing you some interesting examples, you will get on with innovation, to benefit you, your organization and the Asian economies.

We know from the literature on innovation management that, even when one applies all the lessons of innovation management correctly, success requires a dose of luck. So, finally, we would like to wish you good luck in your innovation afforts. May you have much success!

Appendix: List of case studies

Aapico Hitech, Thailand
AmorePacific, South Korea
Asiainfo Holdings, China
AU Optronics, Taiwan
Banyan Tree Hotels and Resorts, Singapore
Biocon, India
ChoLam, Thailand
Dilmah Tea, Sri Lanka
e-Chaupal, India
Giordano, HongKong
Haier, China
Hewlett Packard, Singapore
Infosys, India
iRiver, South Korea
Lapid Foods, Philippines
Li & Fung, Hong Kong
Martha Tilaar, Indonesia
MyWeb, Malaysia, Singapore, China
National Library Board, Singapore
Neowiz, South Korea
Netizen Funds, South Korea
NIIT, India
Patkol, Thailand
Pinoy2Pinoy: SMS in the Philippines, Philippines
Reliance Infocomm, India
Samsung Electronics, South Korea
Samtel Group, India
Semiconductor industry, Taiwan
Shin Satellite, Thailand
Smart Communications, Philippines
Tiger Motors, Thailand
Tata Motors, India
VazBuilt, Philippines

Index